THE WONDER OF ELECTRICITY

THE WONDER OF ELECTRICITY

Kent Mickelson

Copyright © 2023 by Kent Mickelson.

Library of Congress Control Number:		2023913361
ISBN:	Hardcover	979-8-3694-0347-1
	Softcover	979-8-3694-0346-4
	eBook	979-8-3694-0345-7

All rights reserved. No part of this book may be reproduced or transmitted in any form or by any means, electronic or mechanical, including photocopying, recording, or by any information storage and retrieval system, without permission in writing from the copyright owner.

Any people depicted in stock imagery provided by Getty Images are models, and such images are being used for illustrative purposes only.
Certain stock imagery © Getty Images.

Print information available on the last page.

Rev. date: 07/24/2023

To order additional copies of this book, contact:
Xlibris
844-714-8691
www.Xlibris.com
Orders@Xlibris.com
852652

CONTENTS

Introduction ix

Chapter 1 Electricity, the Force 1
    Atomic Parts 1
    Electric Charge 3
    Electromagnetic Force 4
    Electrical Attraction 5
    Summary Comments 7

Chapter 2 Fundamental Electric Concepts 9
    The Flow of Electrons 9
    Electrical Properties of Materials 10
    Voltage, the Reason for Current 11
    Relationship of Current to Voltage 12
    The Concept of Power 14
    Summary Comments 15

Chapter 3 Conduction in Materials 17
    Material Phases 17
    The Gas Phase and Breakdown 18
    The Liquid Phase, the Basis for Solutions and
        Electrolytes 20
    The Solid Phase 20
    Summary Comments 21

Chapter 4	Electricity and Magnetism	23
	Glimpsing the Electricity Magnetism Relationship	23
	A Practical Example of Electricity Related to Magnetism.	24
	A Deeper Understanding	24
	Electromagnetism	27
	Electromagnetic Waves	28
	Summary Comments	30
Chapter 5	Electricity and Light	31
	How Excited Electrons Produce Light	31
	How This Works in Familiar Lights	35
	Summary Comments	37
Chapter 6	Producing Electric Power	39
	Major Types of Voltage Sources	39
	Generators	41
	Voltage Induction	41
	DC Generators	41
	AC Generators	44
	Battery	45
	Solar Cells	48
	Summary Comments	50
Chapter 7	Basic Electrical Parts	53
	Wire and Insulation	53
	Resistors	55
	Capacitors	56
	Inductors	59
	Transformers	61
	Summary Comments	63

Chapter 8 Basic Electric Machines 67
    Summary of the First Part of the Book 67
    Basic Electric Machines 68
        Bells 68
        Relays 71
        Solenoid 72
        DC Motors 72
        AC Motors 74
    Summary Comments 75

Chapter 9 Semiconductor Devices and Applications 77
    P-N Junction Diode 77
    LEDs, Light Emitting Diodes 80
    Bipolar Transistors 80
    Other Semiconductor Devices 83

Chapter 10 The Origins of the Commercial Power System 85
    Components of a Power System 85
    Electric Lights: The First Load 86
    The Light Bulb's Inner Workings 87
    The First Commercial Power System 87
    The Three-Wire Distribution Scheme 88
    U.S. Power System Standards 90
    Summary Comments 90

Bibliography 93
Index 95

INTRODUCTION

It seems like a long time ago that I first gained an interest in the world of electronics. Up until that time, my exposure to electricity was mostly a picture imprinted on my brain by my parents. This came in the form of a fear of what the *electric outlet* represented. For electricity, was, for sure, a dreadful thing, even though it powered all the lights and turned darkness into day?

This fear was planted by the use of a character named Reddy Kilowatt. This character would leap out of the outlet and kill you if you were to put your finger in the socket. I was too young to understand how or why. I'm not even sure that my parents knew. In my mind, I wondered if old Reddy was as real as the Tooth Fairy or even Santa Claus. I was terrified of what might happen if I were to just once disobey the rules. I did not quite associate the battery used in our flashlight with the terror of Reddy. Now that fear has been tempered with knowledge. The fear of the outlet has been replaced by a healthy respect, and I understand now how it is associated with a "simple" battery or a powerful bolt of lightning. The force of electricity is surely a miracle of creation.

Back to my youth, I remember walking the two miles to the local hobby shop. Much to my amazement, there was a little radio in the shop. It was not a kit, but it became an object of my desires.

The storekeeper said that all you needed to do was clip it to a piece of metal, and you would be able to listen to the radio, without even using batteries. I was hooked. However, that steep price of $1.99 was a big obstacle. I only got an allowance of 40¢ a week. I had to put a dime of that into the Sunday school offering plate. However, I was ten years old and just about to get my first job. The $2 every month I got for delivering drive-in movie advertising door to door and the $1.50 a week I got for mowing the neighbor's lawn put me in the big money. I walked back to the store and bought the radio. I was amazed then, and still amazed, even with a degree in electrical engineering, that I could listen to the local radio station without batteries.

What is just as amazing are all the other "miracles" possible when electricity is used. From the beginnings of that tiny crystal radio blossomed the transistor radio, the television, and the modern computer. How can you not be constantly in a state of excitement when you think about the fact that you can use a telephone to talk not only to your local friend, but that you can dial a simple number and talk to your relatives across the country or even on the other side of the world as well?

The applications of electricity in man-made objects are too numerous to describe. However, if you continue reading, let us hope the mystery of how they work will be revealed. This book is meant to help you understand, but it is not meant to help you design. You will gain an understanding of what electricity is and about many electronic components used today. Equipped with those tools, we will look at how those components can be combined into useful things. You will learn the concept of simple design but not be exposed to scary mathematics.

I progressed from the simple crystal radio to the electric motor, the electric bell, the transistor, and then the computer. From what I recall, I learned how to make things that would work long before I understood how they did work. It was not until I was in college that the light bulb of understanding was turned on. My goal is to couple that experimenting and the questions of my youth with the knowledge of my later years to help you. I may not be able to transfer the excitement of the moment I got my first homemade electric motor to work to you, but I will try.

Then too I could not help but be a reminder for the concept of electrical activity in each person. A person might look at the role electricity plays in the beating of your heart, the movement of your finger, the sensation of pain, and the thinking of a thought. I encourage you to investigate this area.

It's all part of the wonderful world of electricity and electronics. I hope you will enjoy it, but most of all, I hope you will learn from it. I hope you will learn to understand the things around you. One spark is all it takes!

CHAPTER 1

ELECTRICITY, THE FORCE

Before we can talk about a definition of electricity, we must take a close look at what makes up matter. Matter is composed of very small particles known as *atoms*. Now you might ask, must we start at such an elementary level? Yes, to understand electricity, it is necessary to understand this concept.

Atomic Parts

Of the many particles that make up matter, an atom is the smallest that cannot be chemically divided. Each atom has a special set of characteristics. Among these are mass and a unique number of protons and electrons that give the atom other physical and chemical properties like color, boiling point, and the ability to conduct heat or electricity. Each different kind of atom is directly associated with an *element*, and the atom passes on its special qualities. All atoms consist of smaller particles, and the combination of these smaller particles gives each element its properties.

Although atoms contain several particles, these are the three basic particles that combine to form an atom:

1. *Proton*

The proton is one of the building blocks of the atom's center or *nucleus*. Although they are thought to be made from smaller particles, protons are considered stable particles. In the context of atomic structure, a particle is defined as stable if it can exist outside the atom without breaking up.

2. *Electron*

The electron is a very tiny particle that moves around the atom. It is stable and can exist outside the atom.

3. *Neutron*

One of the other particles in the nucleus of an atom is the neutron, which has a neutral electric charge, not positive or negative. When more than one proton exists in the nucleus, the neutron and the strong force associated with it keep the protons locked within the nucleus. The neutron's strong force overcomes the effect of individual protons trying to repel each other. The neutron is stable and can exist outside the nucleus.

A normal atom is neutral, which means the number of electrons equals the number of protons. The number of protons and electrons determine the atom's properties.

Each element has a different number of protons. The elements have been organized into a *periodic table*, which presents information about the atomic structure of each element.

The electrons swirling about an atom are put into categories defined as *energy levels*. To help visualize how these levels are structured, physicist Niels Bohr described the atom as being like the solar system, with the nucleus as the sun and the electrons as the planets. Although it is impossible to tell exactly what is happening to a given electron, this concept of energy levels will be useful in later chapters.

The organization of these energy levels, especially the highest energy level, farthest from the nucleus, provides the basis for today's periodic table.

Of major importance when examining electrical properties is the number of electrons in the highest unfilled energy level, and this is called the *valence number*.

Perhaps you're beginning to think we're getting a little off the subject. On the contrary, to completely understand how to approach electricity, the subjects of physics and chemistry are necessary starting points. In fact, electricity plays a fundamental role in particle interactions which allow atoms and matter to exist.

Electric Charge

A basic property of protons and electrons is that they have *charge*. The proton has a positive charge, and the electron has a negative charge. An atom, in its normal state, has a balanced charge. The particles' charges contribute to the existence and properties of the atom.

The concept of electric charge is essential to the understanding of electricity. The force of electricity is ultimately tied to the existence of positive and negative charge. And although we know a lot about what electric charge can do, we don't know exactly what causes it. Even today some college professors might say, "You really can't define what causes charge to exist: That is just the way God made things." We can only accept the fact that matter, as we know it, cannot exist without electrical force.

Electromagnetic Force

For most of this book, we will look at electricity from a macroscopic view, dealing with the consequences of electrically charged particles, not the reason or explanation for the charge itself. Although many texts delve deeply into the interaction between charged particles, we will only take a brief look at one of the widely discussed theories for the interaction of charged particles.

Electromagnetic force is caused by a very special particle, a virtual photon, known as a messenger particle. I like to think of the virtual photon's association with charged particles as being similar to the force that connects the two hands of a juggler. The juggled ball is much like the messenger particle. Some particles juggle, some don't, and some juggle differently. If we could observe the action of two hands as an invisible ball was juggled, what we would see is like the force that holds charged particles together.

Although this analogy is very crude, the intent is to provide a glimpse at the underlying nature of electric force. In any regard, electromagnetic force or the interaction of charge appears to be a direct result of the interaction of charged particles via a messenger particle called the virtual photon.

Electrical Attraction

Did you know that people often rearrange charge without realizing it? Just rubbing a comb across your hair strips some electrons from the comb away from their nuclei, thus their balancing proton, and places them on the hair. When you try to remove the comb, your hair is attracted to the comb. In fact, the property of charge was originally discovered through a situation similar to the comb example. In general, it was discovered that the energy put into a system by work can separate charges.

An atom's structure determines how easily one of its electrons can be removed. Electrons that have been removed are called free electrons. Sometimes an atom can add a free electron to its structure, then the atom is called a *negative ion*. The atom whose electron has been removed is a *positive ion*. These atoms represent an imbalanced charge. By performing work on them, atoms can be put in this "non-normal" state.

COPPER ATOM
(has a neutral charge)

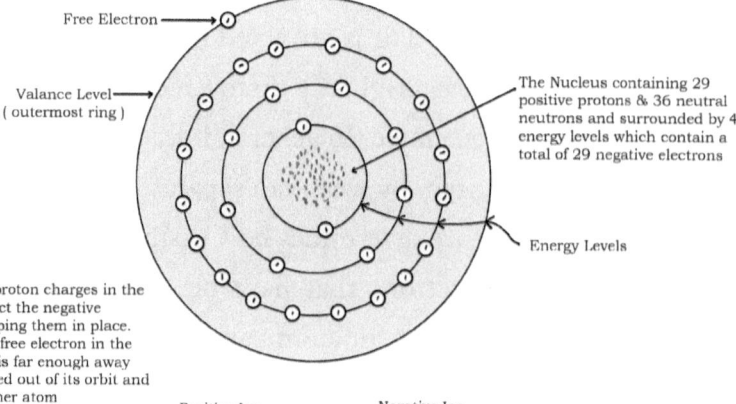

Free Electron

Valance Level
(outermost ring)

The Nucleus containing 29 positive protons & 36 neutral neutrons and surrounded by 4 energy levels which contain a total of 29 negative electrons

Energy Levels

The positive proton charges in the nucleus attract the negative electrons keeping them in place. However, the free electron in the valance level is far enough away it can be pulled out of its orbit and jump to another atom

Positive Ion   Negative Ion

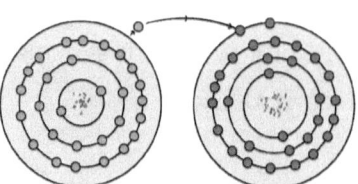

When the free electron jumps to another atom they become ions

We can say that a body's charge is associated with the total amount of unbalance in its atoms. When charged bodies meet, those with like charges repel each other; unlike charges attract. Since the bodies don't need to touch for this to happen, each body is said to have an *electric field* surrounding it associated with its charge.

Coulomb's Law defines the force between charged bodies. It states the force of attraction or repulsion between two charged bodies is proportional to the charge present on both bodies divided by the square of the distance between them.

The units to measure charge are called *coulombs*. The number of unbalanced atoms needed to equal 1 coulomb is extremely large, having twenty digits.

Summary Comments

The main point of this chapter is that particles, and as a result matter, can have an electric charge. Associated with this charge is an electric field that causes particles and/or matter to attract or repel. And it is that force that makes electricity. This principle is used to run everything, from a simple electric light to the most powerful computer.

New Terms

ATOM - The building block of matter. These particles are made of protons, neutrons, and electrons.

CHARGE - The basic property of electricity.

COULOMB - The unit of measure for charge.

COULOMB'S LAW - The law that defines the electric force between charged bodies.

ELECTRON - The very low mass particle that has negative electrical charge and encircles the nucleus. It is the electron that gives substance to the atoms, thus the material by the very large radius around the nucleus in which it might appear.

ELEMENT - The name for matter composed of only one type of atom.

ENERGY LEVEL - The concept used to define the energy associated with an electron. The natural makeup of the atom suggests that this energy can be added only in discrete amounts which then defines each level.

NEGATIVE ION - An atom that has extra electrons.

NEUTRON - This is a heavy electrically neutral particle responsible, through its strong force, for holding the nucleus together by keeping the protons from flying apart.

PERIODIC TABLE - The table that defines the elements based on the number of basic particles that make up its atom.

POSITIVE ION - An atom that has had electrons removed.

PROTON - The particle forming the positive charge of the atom and located inside its nucleus.

VALENCE NUMBER - The number of electrons in the highest unfilled energy level of an atom.

CHAPTER 2

FUNDAMENTAL ELECTRIC CONCEPTS

From the previous chapter, we know that charged bodies are surrounded by an electric field. We also know that the work or energy used, in such activities as combing your hair, causes charge imbalance in the objects. If two bodies have unlike charges, they will attract each other, and that attractive force, as defined by Coulomb's Law, is fundamental to any concept of electricity.

The Flow of Electrons

When a positively charged body is brought into contact with a negatively charged body, the excess electrons from the negative body will move to replace the missing electrons in the positively charged object. Electrons always flow toward an area of positive charge. This "flow of electrons" or discharge is defined as *current*.

Current is measured in *amperes*. The flow of 1 coulomb per second is defined as 1 ampere. A coulomb is a large number: 6.28

x 10^{28} electrons. When the charge of the two areas becomes equal, current flow will stop.

Electrical Properties of Materials

We can change the amount of the current's flow by bridging the space between a positive and a negative charge with different substances. All materials have a certain conductance or resistance when subjected to charge differences.

Materials that allow the flow of many electrons are called *conductors*. Metal wire, for example, is a good conductor of electric current. This is true because metals have a lot of outer-shell single electrons that escape to become free. Sometimes we want to prevent current between two bodies of charge. Materials that do a good job of this are known as *insulators*. Plastic and glass are usually good insulators. When we want a specific amount of current between the current that would flow through an insulator and that which would flow through a conductor, we use a *resistor*. These are usually calibrated slices of a material with well-defined electrical properties. Carbon materials are often used as resistors.

Using conductors and insulators, we cannot completely control the flow of electrons. Because of the characteristic of the space separating an area of positive charge from one of negative charge, sometimes a few electrons can flow between the two areas even though they are separated by an insulator. This is known as *leakage current*.

Static electricity is the term applied to charge at rest or in equilibrium, as described in the comb/hair example. Leakage current and static electricity are the culprits behind the small but annoying shocks you get during winter months. Static electricity can accumulate on your body as you walk across the carpet, an action similar to combing your hair.

Why does this happen mostly in winter? In the summer, the moist air allows a very high leakage current because water increases conductivity, and the charge does not build up enough to cause a noticeable shock. In winter, however, heaters tend to dry out the air, and that decreases the amount of leakage. When the charge on your body builds up high enough, your body will quickly discharge it to lamps and light switches or maybe even other people.

Voltage, the Reason for Current

When work creates areas of opposite charge or a charge difference, there is potential for current flow. This potential is known as *voltage*. A *volt* is defined as the potential difference across an area of charge required to obtain 1 *joule* of work when 1 coulomb flows. One joule is approximately equivalent to the work required to move 3/4 of a pound up one foot. When voltage is present, current flow is possible when two areas of charge are connected.

A voltage can exist without current. A typical battery is an example of the presence of voltage. The chemical work that was done to manufacture the battery produced that voltage. A disconnected battery is a voltage that exists without current flow. When the battery is placed in a flashlight or other device that is turned on, then the battery's two ends are connected, and a current flows. The amount of current that flows depends on the conductance characteristics of the device to which the battery is connected.

The concept associated with a battery is a little different from what was described as static electricity. With static electricity, current flow will usually last only for the short time needed to equalize areas of static charge. A battery, however, will maintain its voltage for some time, although the voltage does eventually disappear. We will take a closer look at ways to create voltage in chapter 6.

Relationship of Current to Voltage

The device connected to the battery is known as the *load*. Its ability to carry current is measured in *ohms*. A special law called *Ohm's Law* states that the current equals the voltage divided by the resistance of the load. A current of 1 ampere (Note: Earlier the actual definition of an *ampere* was discussed under "the flow of electrons") flows when 1 ohm is connected across 1 volt. Ohm's Law, by the way it is written, defines voltage, amperes, and ohms. For example, 1 ampere equals 1 volt divided by 1 ohm. The equation can be adjusted to calculate ohms as well. Conductors have a very small resistance, and conversely, insulators have a very large resistance. The resistance of a material is closely related to the structure of its atoms or molecules.

For example: If you have a battery of 9 volts and a resistance of 1,000 ohms, the current flow is 9/1,000 or .009 ampere, which is 9 milliamperes. There is a series of prefixes that help define electrical terms. A few are

THE WONDER OF ELECTRICITY

micro	– 1/1,000,000 times
milli	– 1/1,000 times
kilo	– 1,000 times
mega	– 1,000,000 times

OHMS LAW

$$V \text{ (volts)} = I \text{ (current)} \times R \text{ (resistance)}$$

amps — ohms

or

$$I = \frac{V}{R} \quad \text{or} \quad R = \frac{V}{I}$$

In the example below we can find the current since we know the volts and resistance

$$I = \frac{1.5 \text{ volts}}{1 \text{ ohm}}$$

$$I = 1.5 \text{ amps}$$

R = 1 ohm
The light is providing 1 ohm of resistance (also caled *load*)

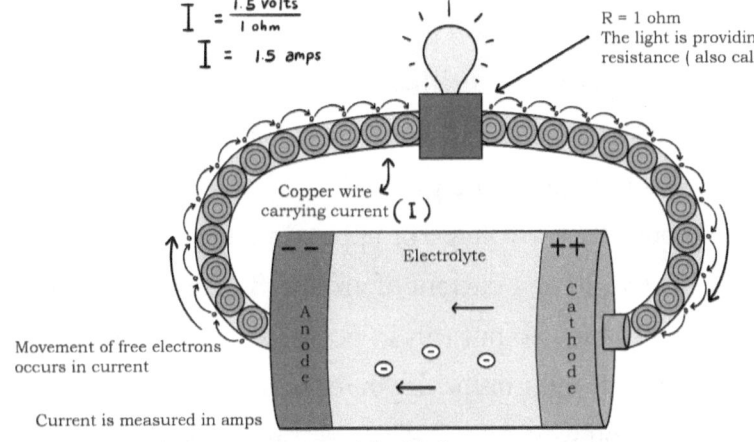

Copper wire carrying current (I)

Electrolyte

Anode Cathode

Movement of free electrons occurs in current

Current is measured in amps

Each atom turns from neutral to positive briefly as its free electron leaves and before the next electron arrives

1.5 volt battery
[also providing 1.5 volts of Electromagnetic Force (EMF)]

The Concept of Power

Just as work is done to produce a voltage, once a voltage is present, it represents a potential to do work. In other words, the voltage represents the ability for *power*. Power does not actually exist until a load is connected to the voltage. *Electric power* should be thought of as the rate at which work is done by moving electrons. Power used by a load can be calculated by multiplying its voltage times its current.

The electric bill sent by your electric power company is usually calculated based on the power or, rather, the work done by the electricity you used. This power is measured in *watts*. One watt is equal to the power delivered to a load of 1 ohm when 1 ampere flows in that load (1 watt can also be defined as a joule/second). Your power bill is expressed in kilowatt (1,000 watts) hours, and you are billed for joules or energy used.

If your house uses a voltage of approximately 120 volts, this is equal to 1,000/120 or a current of around 8.3 amperes flowing for the time of 1 hour. To put this in perspective 1,000 watts will power ten 100-watt light bulbs. If you had these ten bulbs turned on for 24 hours, you would use 2.4 kilowatt hours (kwh). A family might use 500–2,500 kwh per month, more or less, depending on the family and their house.

Most appliances are marked with their power rating. It is easy to determine how much one device contributes to your consumption by multiplying its power rating times the time it is used.

Summary Comments

This chapter has presented some of electricity's basic concepts and terms. In later chapters, we will look at how magnetism and light are related to electricity and how electricity is used to accomplish a variety of tasks.

New Terms

AMPERE - The standard unit of current.

CONDUCTOR - A material intended to allow the flow of electrons across a voltage.

CURRENT - The term associated with moving electrons from higher potential voltage through a closed circuit to a lower potential voltage.

ELECTRIC POWER - The rate at which work is done by moving electrons.

INSULATOR - A material ideally meant to allow no flow of electrons across a voltage.

LEAKAGE CURRENT - The current that flows from one point to another through a material considered an insulator.

OHM - The fundamental unit of resistance as defined by Ohm's Law when Volts = 1 volt and Current = 1 ampere.

OHM'S LAW - The law that states the fundamental relationship between voltage, current, and resistance:

> Volts = Amperes *x* Ohms
> Voltage = Current *x* Resistance

RESISTOR - A material intended to impede or resist the flow of current, usually in a specified way, across a voltage.

STATIC ELECTRICITY - Electricity associated with charge at rest.

VOLT - The fundamental unit of voltage.

VOLTAGE - The term that denotes a difference in charge or a potential for the flow of current.

WATT – The unit of measurement for electric power, can be calculated from the equation Watts = Amperes *x* Volts.

CHAPTER 3

CONDUCTION IN MATERIALS

In the last chapter, we discussed several basic electrical concepts and terms. One of these concepts, conductivity, and its inverse resistivity, is associated with a material's ability to allow electrons to flow. This property determines whether a material is a resistor, substance that impedes the flow of electricity; conductor, material that allows electricity to flow easily; or insulator, substance that blocks the flow of electricity. By using this property, we can confine electricity and control its path.

Material Phases

At this point, we will break material into the three physical phases: gases, liquids, and solids. Water, commonly known in each phase as steam, water, and ice, is a good example for visualizing the differences in molecular or atomic activity of these phases.

The Gas Phase and Breakdown

The motion of the atoms in a gas is sufficient to knock electrons around and create a few positive and negative ions. Gasses are usually considered insulators, but these ions often exist in large-enough quantities to allow a small current when the gas connects areas of opposite charge. A *field* is a pattern of force lines caused by an electromagnetic wave. A field is present when current flows in the gas. As the field increases, the charges can get closer or become bigger, the electrons' motion increases, as does their speed. Finally, an *avalanche breakdown* effect occurs: The current increases rapidly, the gas becomes more excited, the temperature increases, and the energy creates a visible spark. The spark occurs because the magnitude of the field becomes large enough to puncture the insulation.

Air is usually considered an insulator too, but a voltage can generate a field strong enough to cause a breakdown so considerable current can flow. When air acts as an insulator separating charged areas, it is called *dielectric*. All dielectrics have a characteristic breakdown voltage.

Lightning is an example of the breakdown of a gas. During a thunderstorm, a charge difference can build up between the clouds and the ground. That is why a thunderstorm is often called an electrical storm. During a thunderstorm, a tremendous field is created, and the result can be a breakdown of the air that produces what we see as lightning.

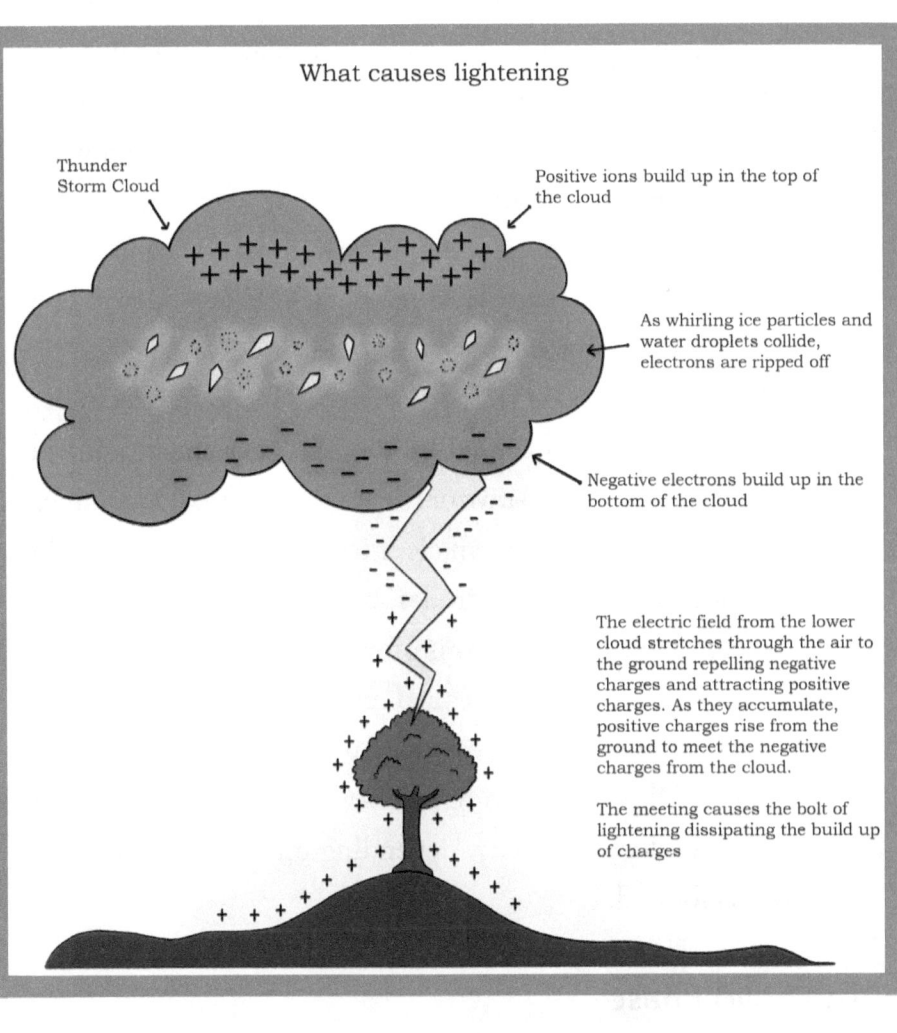

The Liquid Phase, the Basis for Solutions and Electrolytes

Materials in a liquid state are still in active motion and are often very good conductors because they allow the flow of electrons. However, many types of oil are used as insulators, and water, if completely pure, is also a good insulator because they impede the flow of electrons. Water often becomes only a fair conductor when it contains impurities, resulting in a solution containing many positive and negative ions. If the impurity atoms have one or two electrons in the outer energy level or are lacking one or two, they are more likely to become ions.

Although the solution itself appears to be conducting electricity, the transfer of charge is caused by the ions' movement and the chemical reaction at the positive or negative *electrodes*. It is common to refer to the connections to an electrical device, in this case the solution with the ions, as electrodes, one to accept electrons and one to inject them. This process is often called *electrolysis*, and the solution in which it occurs is called an *electrolyte*. This property leads to the creation of some very useful kinds of devices we will discuss later. All you need to remember now is that liquids may or may not conduct electricity, depending on whether a solution with ions is formed.

The Solid Phase

Solids make up the bulk of the commonly known good conductors. However, some solids, such as glass and plastic, are very good insulators. A material's chemical structure is the main factor in determining its electrical properties. Most solid insulators are made from complex molecules that do not involve atoms that have loosely bound electrons waiting to be ionized.

On the other hand, most metals have a regular structure involving atoms with loosely bound electrons, which makes metals good conductors. Irregularities in the structure of the metal will decrease the conductivity. As the temperature of a metal increases, thermal vibration causes electrons to scatter. The effect of this reduces conductivity. Even in a good metal conductor, the motion of the electrons causes the structure to vibrate and temperature to increase.

Summary Comments

The number of *valence electrons* of a material's atoms dictates whether a material will be an insulator or a conductor. The valence electrons are the outer-shell electrons that tend to produce the properties of an element. This factor determines the number of ions created, and as a result, it defines the quantity of loose electrons, once the material's basic properties are defined, they are valid only within finite limits. An insulator can break down and conduct, or under a high current, a conductor might gain more resistance. Failure to respect those limits has led to many electrical fires by causing high temperatures through conduction or even short circuit.

New Terms

AVALANCHE BREAKDOWN - The point at which an insulator fails to insulate but instead conducts.

DIELECTRIC - Another term for an insulator that separates two charged surfaces.

ELECTRODE - A conductor by which current flows into and out of a substance, often a fluid.

ELECTROLYSIS - The flow of current into a solution, causing its chemical composition to change.

ELECTROLYTE - A solution whose chemical composition is changed by action of an electric current.

FIELD - A region under the influence of electromagnetic particles.

SOLUTION - A fluid that contains dissolved impurities.

VALENCE ELECTRONS - The outer-shell electrons of an atom.

CHAPTER 4

ELECTRICITY AND MAGNETISM

Magnets are well-known objects to most people, and they are used throughout most homes. Think of all the magnets on refrigerator doors, for example. A discussion of magnets in this book is important because their existence is closely related to electricity.

Glimpsing the Electricity Magnetism Relationship

Magnetism manifests itself as a field, which has an inverse square relationship. The strength of the field decreases according to the square of the distance to the source. A magnetic field is produced by a current or, rather, a moving charge. In addition, researchers have determined that a moving magnetic field can provide the force to put a charged particle in motion. In a similar manner, when an electron charge is moving through a stationary magnetic field, a force is exerted on the electron. We can say, if the magnetic field is moving in relation to a charged particle, a force is exerted on the particle.

A Practical Example of Electricity Related to Magnetism.

In a television picture tube, a precisely controlled magnetic field is used to guide a stream of electrons to the surface of the screen. The inside surface is coated with a special substance that emits light when electrons hit it. The television signal varies the intensity of the electron stream. Every TV is made with certain assumptions about the control of the electron stream so that a picture will appear. As long as the receiver and the transmitter make the same assumptions, a meaningful picture is seen. Although this is a very brief description, it gives you a concrete picture about a well-known device that takes advantage of the relationship between electricity and magnetism. There are many devices that do this. For example, motors, generators, and radios also take advantage of this relationship.

A Deeper Understanding

To really understand magnetism, we must once again refer to matter at the atomic level. The magnetic properties of all materials arise from their molecular structure. A simple overview of the atom is that it consists of a very small nucleus, with electrons orbiting around that nucleus. Although it is not possible to define the path of a single electron, with a statistical view, we can imagine the electrons orbiting the nucleus in a somewhat circular fashion. In other words, we can view the electron as a tiny current flowing about the nucleus. We can also view the atom as a miniature solar system, with the nucleus as the sun and the electron as a planet.

Don't take this analogy too far, however, because the electron orbit is much more random in nature than a planet's path.

An atom becomes a small magnet because of the magnetic field produced by the tiny current its orbiting electrons create. The spin of the electron itself also produces additional magnetic effects. In fact, the motion of the charged mass of the electron spinning produces a larger magnetic effect than its orbit around the nucleus. In multiple-electron atoms, some of the orbiting negative electrons pair up and cancel their magnetic effects. This means the number of electrons in an atom, as well as their configuration, can affect the total magnetic field strength of the atom.

Wherever a moving electric current exists, so does a related magnetic field, and this produces the magnetic properties of any given material. With a closed circle of current, the result is a magnetic field with the well-known north and south poles; a magnetic field will always exist with a positive and negative ends to the field. For example, when an electron orbits around a nucleus, or an electron itself spins, or a current flows in a loop of wire, this closed circle of current exists. A magnetic field has lines of force called *magnetic flux* that connects the positive and negative poles.

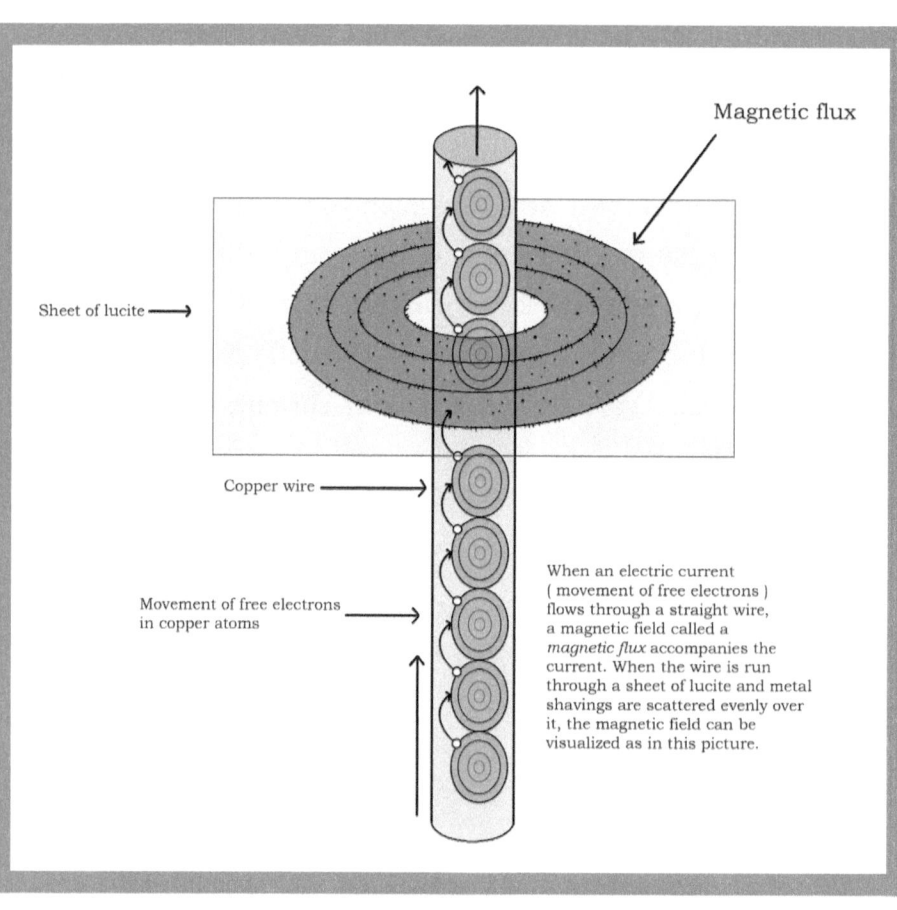

An atom displays magnetic effects because of the orbit and spin of its electrons. The orientation of the atoms in the structure of a material is what determines the magnetic characteristic of that particular kind of material. Some materials have no tendency for magnetism at all. Others will orient themselves in magnetic alignment only in the presence of a magnetic field from another source. In other words, many materials have atoms that are individually magnetic, but the atoms are so naturally disorganized that the material itself does not act as a magnet until its atoms are organized, aligned, by an outside force. Some materials, once aligned, will remain aligned and be permanently magnetized. Others will retain very little magnetic properties once the external field is gone. The magnets you use to fasten things to your refrigerator are permanent magnets that were made using this knowledge. These permanent magnets will only attract materials that will align themselves to their magnetic field.

Electromagnetism

An important application of these effects is the creation of an electromagnet. For example, if you wrap an insulated wire around an iron nail several times (Note: the wire must be insulated so the effect of many current loops is created) and attach a battery to the ends of the wire, the resulting current will induce the atoms of the nail to align, and an electromagnet is created. The magnetic field produced by this effect produces what is called *electromagnetism*. Once this is done, the properties of the material will determine whether it will remain aligned after the battery is removed. For many objects, the fact that the alignment is not permanent is a very important property. For example, electrical machines like motors depend on the ability to reverse the current of an electromagnet and force the electromagnet to quickly reverse its polarity.

If you take that same nail and wire but without the battery and move it in the proper way through a magnetic field, a measurable voltage is produced between the ends of the wire. These complementing processes are the foundation for the operation of many types of electric machines. They are a direct result of the interdependence of magnetic and electrical forces, which are unified in what has become known as *electromagnetic theory*. The forces associated with current and magnetic fields can be remembered very clearly by something called the left-hand rule. If you point your left thumb in the direction of a moving charge, the curl of your other fingers provides an idea about the direction of the associated magnetic fields.

Electromagnetic theory involves a lot of complex mathematics; therefore, we won't delve into it in great depth. The basic concepts discussed in this section give an understanding of the theory's affects, and for our purposes, that is enough.

Electromagnetic Waves

There is one implication of electromagnetic theory, however, that needs to be mentioned in greater detail: the wave concept. The motion of electrical energy is often associated with waves called *electromagnetic waves*.

You've probably heard of radio waves and light waves. They are forms of electromagnetic waves. For all practical purposes, electromagnetic waves are a way in which energy is transmitted through space, even through a vacuum. It's a way in which the electric field at one instant creates a magnetic field perpendicular to it, and the resulting magnetic field creates an electric field. More precisely, this relationship is determined by a set of equations defined by the famous *Maxwell's equations*. These state that a changing electric field creates a magnetic field linking the electric field. But this magnetic field is also changing, and it produces an electric field that sort of forms loops about the magnetic field. The relationship between the fields propels forward the energy within the fields. In outer space, the speed of this propagation is equal to the speed of light. In a simplified way, this is a description of what an electromagnetic wave is.

Certain experiments show the wave nature of electrical energy. Other experiments, however, demonstrate that the same waves can be interpreted as particles. Light waves are a common example. Some experiments show the presence of particles called photons, while others reveal the creation of interference patterns associated with waves.

The point of this discussion is that electromagnetic theory is what predicts and allows us to deal with electromagnetic waves, and these waves affect the lives of every person on earth. For example, the life-giving energy of the sun would never reach the earth without electromagnetic waves. In fact, this theory of electromagnetics is so intertwined with the microscopic and the macroscopic world that it provides the glue for chemical bonding and also the passage for the energy that allows life on earth.

Summary Comments

We have taken the first step toward learning the fundamental theory of much of electrical engineering by discussing electromagnetic theory. I have tried to show you the effects of this theory rather than try to develop a detailed explanation of the theory itself, which would require some rather overwhelming mathematics. Of great importance is how electricity and magnetism are related, as demonstrated in the example of the nail with wrapped wire.

New Terms

ELECTROMAGNETIC THEORY - The theory that defines the relation between electricity and magnetism.

ELECTROMAGNETIC WAVE - The oscillating transmission of energy as defined by electromagnetic theory or the perpendicular relationship between the electrical and magnetic field.

ELECTROMAGNETISM - The result of the magnetic field created around a current caused by the flow of current through it.

MAGNETIC FLUX - the lines of force connecting the positive and negative poles of a magnetic field.

MAXWELL'S EQUATIONS - The fundamental set of equations that relate electric and magnetic fields in mathematical terms.

CHAPTER 5

ELECTRICITY AND LIGHT

Light is an interesting aspect of electricity because almost everyone can relate to it. When we see the colorful rainbow of a spectrum of light, it's a concrete reminder of how electrons and electricity work.

Most people know electricity must, in some way, be related to light. Beginning with Thomas Edison, people have grown accustomed to a variety of electric lights. This may be one of the most important things we can attribute to electricity.

How Excited Electrons Produce Light

Once again, to see how electricity affects us, we must look at the atomic level. If you recall, we set up a model of the atom with protons in the nucleus surrounded by swirling electrons. The electrons travel in orbits, which we associate with energy levels.

By pumping energy, such as electricity, into an atom, the atom becomes "excited," and its electrons jump to higher energy levels. Later the electrons fall back to their lower, "normal" energy levels, but as they do this, they release energy in the form of a photon or particle of light. This is called released radiation.

The increments in which these events happen have resulted in a whole field of study called *quantum mechanics*. Scientists noticed that energy is added or subtracted from an atom only in distinct intervals, quanta. The energy-exchange occurring with energy-level quanta was further developed by physicist Max Planck. *Planck's equation* states that the energy of this quantum of light is equal to a value called Planck's constant times the frequency of radiation. What this equation does is to relate the quantum energy of an electron dropping in energy levels to the creation of light or other radiation.

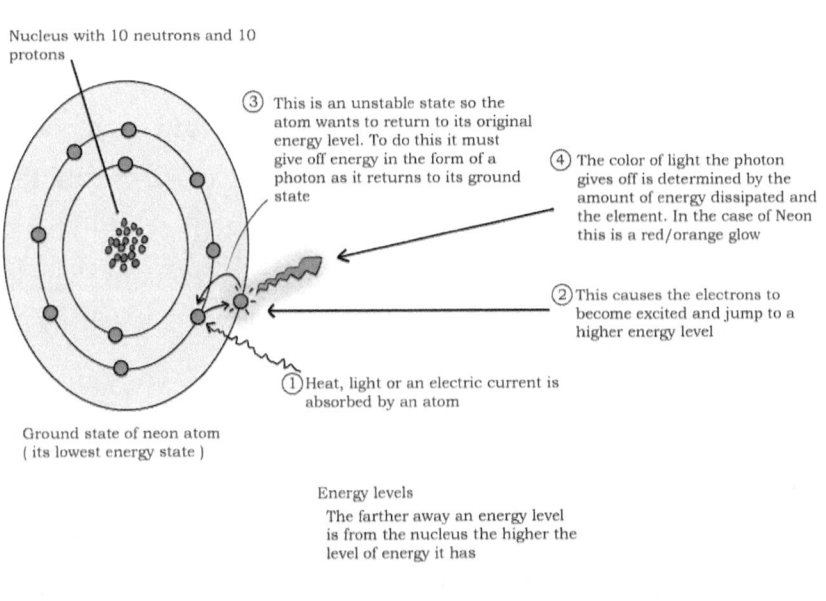

One of Planck's interesting findings is that the light's color is caused by the amount of energy in the quantum and the frequency of the emission. As an example of how this works, let's look at the simplest atom, hydrogen. If you excite a tube of hydrogen gas by passing an electric current through it, you see a light. If the light passes through a prism, you would see distinct color lines instead of a continuous spectrum of light. These lines are known as the *atomic spectra*. Using the concepts provided by Planck's equation, this spectrum can be predicted and understood.

So what happens to make these different colors? Suppose an electron's normal state is in level 2. When it gets excited, it can zoom up to any number of higher levels. Accordingly, the "normal state" the electron falls to may be any lower level, at which point a quantum of energy is released. The radiation released by different falls, quantum energies, will produce a different color. That is the reason for the atomic spectrum. This is all related to Planck's equation which states the frequency of emission is equal to the energy in a quantum of energy divided by Planck's constant.

When you look at a bunch of atoms, the number of times each atom is excited, so it subsequently gives off a photon, produces the average color for a tube filled with that kind of atom. Therefore, hydrogen gaslights give off a different color from neon lights. Each would display an atomic spectrum different in either lines or magnitude. To explain differently, a particular atom will have atoms that have electrons in orbit that can absorb energy, become excited, and jump to a higher energy level and later fall back down to a lower level, giving off a photon of light defined by Planck's equation. This process gives off a characteristic color for each element. Often this can be seen when an element is burned. In this case, the burning adds energy to the atoms. Each element will usually produce a characteristic color producing distinct patterns in a spectrometer, a device for dividing a color into wavelengths allowable by possible quanta. In fact, scientists examining these lines discovered that atoms have a more complex structure than they originally thought.

When an electron drops to a lower quantum level, the lost energy causes the emission of a photon, which might be seen as visible light. And the ingredient responsible for knocking atoms around so that they emit light is an electric current.

How This Works in Familiar Lights

In this chapter, we will talk about three kinds of electric lighting devices that use this concept: fluorescent lights, computer/TV screens, and incandescent lights/light bulbs.

Fluorescent lights are actually tubes filled with gas. If you examine the tube, you'll see metal contacts at each end. As an electric current is forced through the gas, the process we described begins to take place, producing some visible and much ultraviolet radiation. A fluorescent material inside the tube is excited by that radiation and produces the light we see.

In older TVs, the picture is displayed with a cathode ray tube. This is also the same for early computer screens. In these tubes, the screen, which is the picture, is created by a controlled stream of electrons using what is called an electron gun, a directed beam of electrons, toward the inside of the tube. This beam contains intensity information in the signal from the TV station. The electrons hit a phosphorescent coating, and those excited electrons make the light we see, with different coatings producing different colors. Quite often color is derived from a repeated pattern of red-, green-, and blue-type coatings in various combinations. The signal causes the electron gun to sweep from the top to the bottom in tiny increments so that each spot on the screen receives information only a portion of the time. This was also discussed in the section on electricity and magnetism.

The last example is the "lowly" incandescent light bulb. This is the light that really brought the magic of electricity onto our streets and into our homes. The electric light changed people's lifestyles forever. These bulbs first made their appearance in the 1800s, and many innovations have gone into them ever since.

Basically, a light bulb involves a current flowing through a filament, and as the filament becomes hot from the molecular motion, excited electrons start producing light and significant heat. Producing electric light also requires some things discussed before: a voltage, a load made of all things to which electricity is delivered, and the current that flows. In the example, electricity flows into a load made up of a conductor to channel the electricity and the light bulb itself; in other words, a closed loop for the "flow" of electrons. In the rest of this book, this type of arrangement will be described as an *electric circuit*.

Light can be produced in other ways as well. The most common is associated with burning things. With burning, heat is the energy source rather than electricity. In today's world, much of what we see is the LED bulb. We will address that later in chapter 9.

Summary Comments

This chapter has been devoted to describing how electricity is related to light. We have seen that the stimulation of an atom can cause electrons to become excited, and they later fall into a lower energy state that causes the release of a quantum of energy. This may be seen as visible light caused by a photon's release. Planck's equation is important in this chapter because it explains how different colors of light can be emitted.

This chapter also examined some common sources of light and explained how that light is produced. The history of our electrical system reveals that the ability of electricity to stimulate the production of light caused the growth of the electricity network we see today.

New Terms

ATOMIC SPECTRA - The frequencies of light emitted from electrons falling from one energy level to another, which may create a photon.

ELECTRIC CIRCUIT - A closed loop of electrons, current, caused by a voltage being connected to some sort of load.

PLANCK'S EQUATION - The equation that sets a quantum of energy equal to the frequency of radiation multiplied by a special constant labeled Planck's constant.

QUANTUM MECHANICS - The field of study that examines energy in distinct units.

CHAPTER 6

PRODUCING ELECTRIC POWER

In this chapter, we will investigate the major ways continuous voltage is produced. We will discuss the most common sources, which involve electromagnetism, electrochemistry, and light, photo voltaic. You should be aware, however, that voltage can also be generated by pressure, friction, and heat.

Major Types of Voltage Sources

Before we start looking at voltage in any detail, we must begin with the distinction that continuous voltage can be *direct current* (DC) or *alternating current* (AC). The voltage at the electric outlets of your home is classified as AC, while common flashlight batteries have DC voltage.

Direct current voltage stays constant, whereas alternating current voltage varies over time. AC voltage is produced with a characteristic cycle, and if we were to continue to observe this cycle, it would repeat over and over. The time it takes for one cycle to take place is the *period*. Often AC is referred to by its *frequency*, the number of cycles happening each second. This is often seen as cycles per second (CPS) or *hertz* (Hz), in honor of Heinrich Hertz, who was the first person to broadcast and receive radio waves. Technicians use an instrument called an *oscilloscope* to view images of voltage versus time.

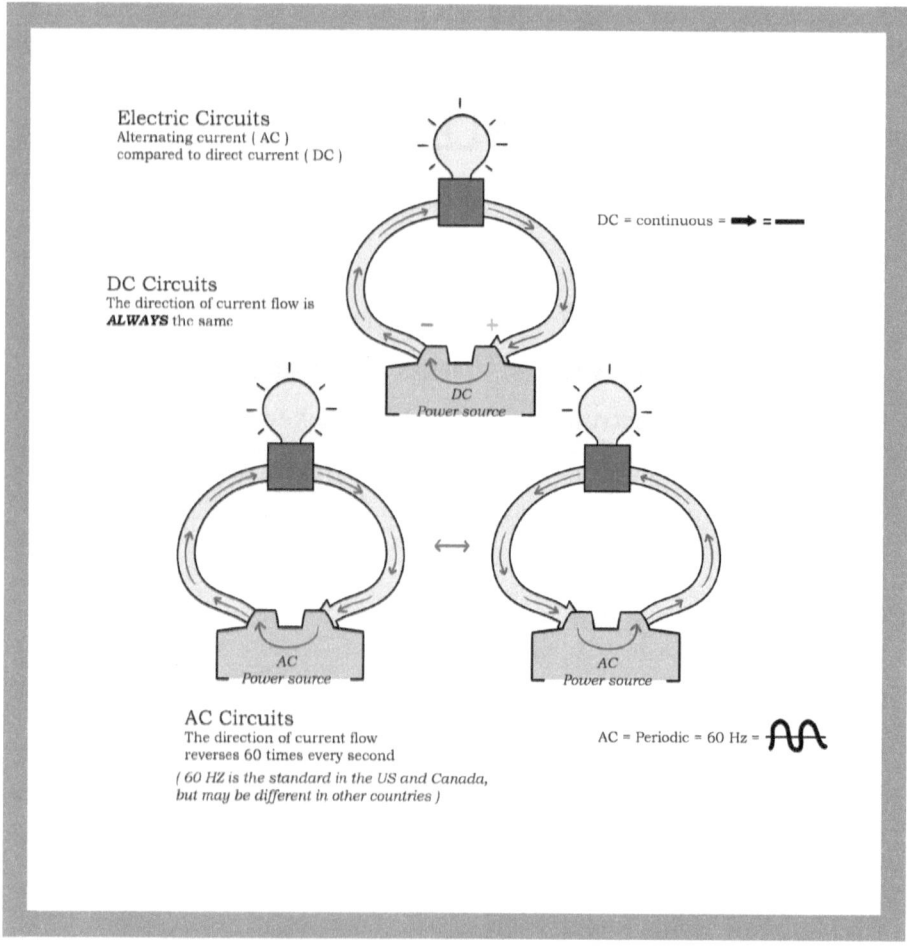

Generators

Voltage Induction

A measurable voltage will occur between the ends of a conductor when the conductor is moved in the "right way" through a magnetic field. Introduced earlier, a magnetic field has lines of force called *magnetic flux* that connects the positive and negative poles. Their density is proportional to the strength of the field. Whenever the flux lines crossing the path of a conductor change, a voltage is induced in the conductor. This change can be caused by the motion of the magnetic field itself, or it can be caused by the motion of a conductor within the magnetic field. This is known as an *induced* voltage, and it's the key concept in making electricity generators.

DC Generators

DC and AC voltage can be created using generators. In a DC generator, there is a field, an *armature*, a *commutator*, and *brushes*. Note that the field can be created by a fixed permanent magnet or a wire winding.

As the armature is turned, the presence of the magnetic field induces a voltage in the coils of the armature. The commutator passes current from the turning armature to the brushes. The electrical connections made via this system keep the generator's voltage at the same polarity for the entire revolution. Polarity relates to the direction of current flow caused by a voltage. The rotation relates to the output voltage. Note, particularly, that the maximum voltage occurs as the coil is cutting through the most magnetic flux.

To change the voltage output, the number of loops in the armature winding can be increased. In addition, increasing the rotation speed causes the voltage output from the armature to increase.

In a DC generator, several windings can be incorporated at various angles to make the actual voltage come closer to ideal, constant. This happens by splitting up the commutator into more pieces and connecting the different windings to the commutator's separated parts. As it is, most DC generators have a slightly varying DC, which is defined as *voltage ripple* content. Many modern generators consist of complex multiple coil armatures with associated multiple contact commutators.

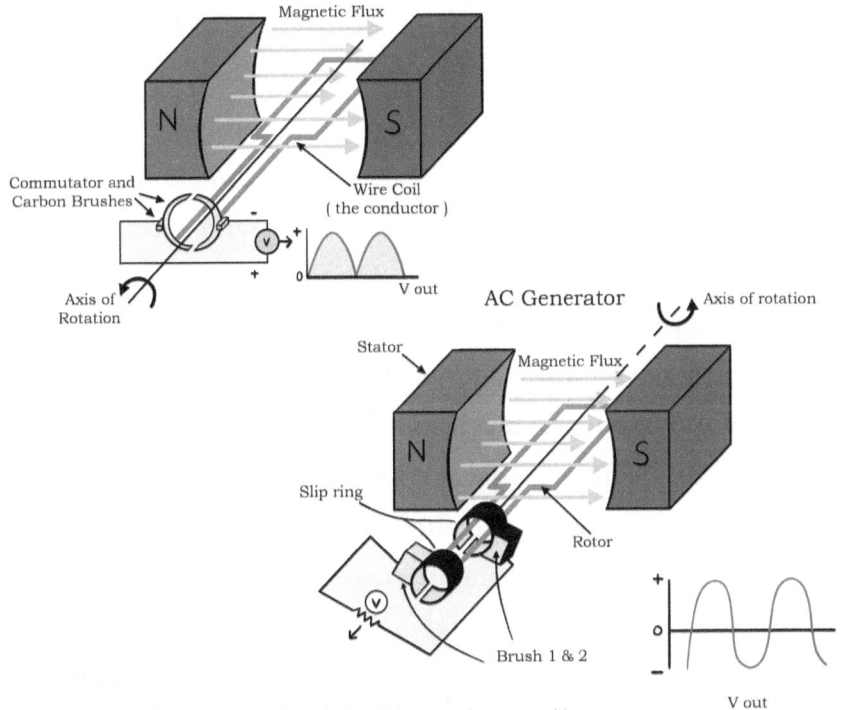

AC Generators

Generators are also made to create AC voltage, and the basic concepts are like DC generators. But AC generators are described by additional terms not used in the DC counterpart. Although it still has a similar armature and field, an AC generator can be set up with either a revolving field or a revolving armature. The conductors that generate the voltage are called the armature, and the conductors that set up an electromagnetic field of constant polarity are known as the field windings.

To make identifying parts easier, the rotating part is known as the *rotor* and the fixed part is known as the *stator*. Rather than conducting current to the rotor via brushes and a commutator, an AC generator transfers a continuous current through two unbroken slip rings to the brushes. Each end of the coil is attached to one of these slip rings, and one brush is assigned to each ring. This method provides a much smoother transfer of current when compared to the multiple-piece commutator of a DC generator.

In the common DC generator, voltage is induced in a rotating armature. But in its AC counterpart, a common scheme is to apply an AC voltage to the rotor, and as it turns, the poles passing by the fixed windings induce the voltage into the fixed windings. As the rotor turns, an alternating voltage of the same polarity is generated at the output.

The armature winding of the AC generator can also be arranged so more than one voltage *phase* can be created. Phase is related to the time relationship between one voltage waveform and another.

Certain types of loads often require AC generators with multiple phases. We've already seen how single- and two-phase systems are created, but three-phase AC power systems are also in wide use. As we discuss how these voltages are used, you'll see why these multi-phase generators are necessary.

It's important to remember the electricity generated does not come from the magnetic field; it actually comes from the mechanical energy input by turning the coil. In a generator, the coil is often turned by the force of water, steam heated by coal or nuclear energy, or gasoline or diesel engines. Nevertheless, the magnetic field is essential—a voltage can be present only if work is done to separate charges. This potential voltage is present only when a generator's rotor is turned, whether by engines, water, or wind.

Here are three examples from our daily experience:

1. Very large - Generators at power plants
2. Medium - An automobile generator
3. Small - A bicycle light generator

The AC voltage used in homes usually comes from large generators at a central location. Later, we will take a closer look at how that power is generated, transmitted, and distributed to our homes and businesses.

Battery

To begin talking about batteries, we need to talk about a few terms. We think of batteries as consisting of multiple cells connected. A *cell* is a device that converts chemical activity into a measurable voltage. An *electrode* is the conductor that enables the electrons to leave or enter the battery. *Electrolyte* is the chemical solution where the electrodes are placed.

Two types of battery cells are *primary cells* and *secondary cells*. In a primary cell the chemical action uses up one of the electrodes. One significant problem with primary cells is that they wear out irreversibly. Secondary cells, despite functioning similarly to primary cells, can be recharged. That is why they are often called *storage cells*. This type of battery stores the energy as a chemical reaction.

One of the simple primary cells is the *galvanic cell*. In this cell, the electrode materials are zinc and carbon, and the electrolyte is a solution of some nasty stuff known as sulfuric acid. This electrolyte contains ions of sulfate and hydrogen, with the sulfate being negative and the hydrogen positive. When the electrodes are placed in the electrolyte and a wire is connected between them, a current will flow.

Why? The flow occurs because the sulfate ions combine with the zinc electrode, and the hydrogen collects at the carbon electrode. The action of the ions at the electrodes produces voltage. The amount of voltage available between the electrodes depends on the materials used to make those electrodes. The simple cell in this example has approximately 1.5 volts.

It's important to remember the cell itself has an inherent *internal resistance*. In other words, as the current supplied gets higher, the voltage between the electrodes decreases. The size of the electrodes and their spacing influence the amount of internal resistance. As the chemical reaction eats away at the electrode, the galvanic cell wears out because the number of ions and the amount of electrode decreases.

Although it has been a simple look, this is the basic process that occurs in primary cells. The materials may change, but the process is always similar.

Some primary cells differ in that their electrolyte is more of a paste than a liquid solution. They are called dry cells, although if the paste were to lose its moisture, the cell would not function. This paste usually contains a variety of materials, but the main ingredient is some form of electrolyte.

An example of this type of cell is the ordinary flashlight batteries, which are often zinc-carbon dry cells. The zinc electrode is the can of the battery, and the carbon rod with a metal end cap is suspended in the cell. As the cell is discharged, the chemical action may cause the cell to expand. If nothing encapsulates the zinc can, the infamous leaking battery will result.

One of the most common secondary or storage cells is the lead-acid car battery. This battery consists of six separate cells in a series. If current is forced into the cell in the reverse direction of its discharge, the cell can be restored. Because the chemical makeup of the electrolyte and both electrodes change during discharge, that change can be reversed during the charge process. The result is that, with periodic charging, we have a continuous source of voltage.

One of the effects of the charge current, however, is the production of hydrogen gas. That is why those batteries require some degree of caution since the air/hydrogen mixture is quite explosive.

Besides the well-known lead-acid battery, other familiar materials are used for storage batteries. One of these is known as the lithium battery. Another example, the nickel-cadmium combination, which is common and far superior to the lead-acid battery. Both battery types play a major role in the ways all of us use electricity.

Solar Cells

To discuss solar cells, we first should talk about some additional topics concerning conduction in materials. Earlier we discussed one class of element characterized as being neither a good insulator nor a good conductor. They have become known as *semiconductors*, and they are the elements on the periodic table that have a valence number of four, such as germanium or silicon.

Elements that have a valence number of one tend to be good conductors; that is, they easily yield a free electron. The greater the valence number, the more difficult it is to free the extra electrons in that high energy level.

During our discussion of electricity and light, we found that light is generated when electrons in an atom fall from high energy levels to lower energy levels. As this change occurs, a quantum of light called a photon is created and sent out in a color related to the size of the transition. In reverse, photons that hit any material cause the electrons to get excited, which often contribute enough energy to some atoms that an electron is freed. The intensity of the photon's light affects this activity, and the materials that exhibit this effect the most are the semiconductors.

As light hits a semiconductor, like silicon, many atoms become so excited that a lot of free electrons are formed. As soon as they are formed, however, they immediately fill in empty spaces in other atoms. The atoms that give off an electron are commonly referred to as *holes*. In a pure semiconductor, as soon as a hole is made, a free electron quickly pops in to fill it. A clever way to take advantage of this effect involved the creation of two new types of material. Other elements having either three or five valence electrons are combined with the silicon in a process called *doping*. With boron, valence 3, doping yields a crystalline structure that gives boron the tendency to gobble up free electrons. With phosphorus, valence 5, doping yields a crystalline structure that tends to yield more free electrons than before doping. In other words, there will be more holes in the first combination and more free electrons in the second. These are known as P-type and N-type material, respectively. To make a solar cell, also called a *photo-voltaic cell*, a thin layer of N-type material is bonded to a layer of P-type material. As photons hit the N-type material, many free electrons are produced, which remain free because of the doping material. If a wire is attached between the N-type and P-type materials, a current will flow as electrons move to the area of excess holes.

The electrons will continue to cross over the boundary area, but eventually, a neutral cell barrier is formed when the closest holes are filled. The rest of the free electrons are available to produce a current. The surface area is related to the amount of current available; however, to produce a higher voltage, they must be connected in series. (Illustration for semiconductor on page 79)

Summary Comments

In this chapter, we've discussed the three major ways voltage is generated today. We have learned:

1. How mechanical energy uses the basic concepts of electromagnetic energy to produce AC or DC voltage through generators.
2. How chemical processes are harnessed to make batteries.
3. How sunlight is captured and turned into power by a simple method using semiconductors.

Now that we know how a voltage is produced, the following chapters will discuss different ways voltage is used.

New Terms

ALTERNATING CURRENT VOLTAGE - A voltage that varies over time, characterized by a repetitive pattern.

CYCLE - A complete pattern of an AC repetitive pattern.

FREQUENCY - The number of cycles per second of an AC voltage.

INDUCED VOLTAGE - The voltage that appears in a conductor as a result of moving it through a magnetic field.

MAGNETIC FLUX LINES - Lines of force connecting positive and negative ends of a magnetic field.

OSCILLOSCOPE - A measuring instrument used to examine how voltage/current varies with time.

PERIOD - The time necessary to complete one cycle of an AC voltage.

DC GENERATORS:

ARMATURE - The coils connected to the rotating shaft. This is the part of the generator that cuts through the magnetic flux lines as it turns, creating a voltage.

BRUSHES - The part of the motor that contacts the rotating commutator, allowing the voltage on the armature to be used.

COMMUTATOR - The connection through which the armature voltage leaves the rotating shaft. It's also the means for converting the generated voltage to DC.

AC GENERATOR:

ROTOR - The rotating portion of the generator, usually with a fixed magnetic field causing an EMF in the stator as it rotates.

STATOR - The fixed part of the generator where the generated voltage usually originates.

BATTERY:

CELL - A device converting chemical action into measurable voltage.

ELECTRODE - Conductors that allow electrons to enter or leave.

ELECTROLYTE - The chemical solution where the electrodes are placed.

PRIMARY CELL - Cell type in which only one electrode is altered.

SECONDARY CELL - Cell type in which reversible processes occur at both electrodes.

SOLAR CELLS:

DOPING - Adding "impurities," combinations of other elements, into a semiconductor.

HOLES - A semiconductor that has given off one electron, leaving a vacant place for an electron.

N-TYPE MATERIAL - A semiconductor that has been doped with a valence number 5 material.

P-TYPE MATERIAL - A semiconductor that has been doped with a valence number 3 material.

SEMICONDUCTORS - Elements having a valence number of 4.

CHAPTER 7

BASIC ELECTRICAL PARTS

This chapter contains the practical aspects of insulation, wire, resistors, capacitors, inductors, and transformers. Our goal is not to make these parts recognizable, in the physical sense, but primarily to gain an appreciation for how they work and what they are used for. I think this is a good first step for gaining insight into why they came to be and put us on the road to removing some of the mystery of how to use electricity. As we discuss the devices, keep in mind each device, including wire, can be described in a symbolic way to help define a circuit representation called a *schematic*.

Wire and Insulation

Wire and insulation are the cornerstones of electrical devices. They are so intertwined that neither has a function without the other. Wire, which consists of conductors and insulators, is the material that channels electricity. Although wire is made from other conductors, it's usually made from copper. There is a wide variety in how wire is made and what its diameter is. In a schematic, a wire is represented by a simple line. A way of forming a wire is also created by making a copper path on a *printed circuit board*. Printed circuit boards were invented to enable manufacturers to replicate a circuit in a more efficient and reliable way.

Usually, wire is made with a center conductor surrounded by an insulator. Insulators help route/contain electricity flowing in wires, which is very important in preventing harm to people when voltages get high. Insulators are also crucial in keeping different voltage wires from short-circuiting, touching one another, together. Sometimes the air itself serves this later function, allowing otherwise uninsulated wire to be used without the threat of harm. Such situations are approached with caution in case dangerous voltages might be exposed.

The diameter of the wire is related to the *cross-section area*, the area of the end of the conductor after it is cut in two. This area is directly related to the resistance of the wire. As larger currents begin to flow, heat will build up because the wire is dissipating energy. If too much current begins to flow, the heat can build up so much that it could become a fire hazard.

Because of this potential for hazard, wire has been classified by size. Wire conductor size is normally defined as a *gauge*. The larger the gauge, the smaller the conductor diameter. For an idea of size, extension cords for use in and around the home are usually 14–18 gauge.

Another way wire conductors differ from one another is in the number of strands used in their construction. A wire conductor cross-section is built of one strand, called solid, or many smaller strands, called stranded. There are two styles for physical and technical reasons. For example, when the current being passed through the wire is made of high frequency AC components (stereo cable), it is often better to use stranded wire, not only because of flexibility, but also because the wire characteristics concerning electricity flow are better. In other words, the number of strands determines the wire flexibility and the characteristic the wire has in conducting higher frequencies.

The insulator is also characterized by the amount of voltage it can withstand before a certain amount of leakage or "breakdown" happens. In some cases, set standards specify requirements for a wire's size and insulation. The usage of a wire determines the selection of size and insulation and the knowledge of the way a particular conductor and its associated insulation function.

Resistors

Resistors are devices specifically intended to limit the flow of electrons across a voltage. Resistors are available in two main commercial package styles. For small resistors, usually made of carbon, the resistor's size indicates the *power dissipation* ability. The smaller the resistor, the less its capacity to dissipate power. Power dissipation refers to the energy that can be absorbed as heat as the product of V x I (voltage times current). For many resistors, the resistor value in ohms is color-coded on the case. Knowing the resistor color code removes some of the mystery when you look at an electronic device. The most common resistors are ¼ and ½ watt. The resistors are also marked to identify the accuracy of their value, for example, 5 percent or 10 percent.

Besides carbon resistors, there are other substances used to make resistors. They can be a solid core resistor or a film resistor.

However, the other major type discussed here is the larger power resistors made of high-resistance wire wound on a core and encapsulated in an insulator. These are large enough that the power rating and value are stamped on the case. There are also military-grade resistors which have less tolerance.

Resistors are also made to be varied by sliding one contact in either a straight line or a circle across the resistive element. Such a variable resistor is known as a *potentiometer*, if the resistance element is carbon, and *rheostat*, if the resistance element is wire. Potentiometers are often used as volume controls for amplifiers.

Capacitors

Capacitors are elements that store energy by sandwiching a dielectric between two conducting plates and attaching terminals to those plates. Many types of package styles are available, in part, because the dielectrics may be air, paper, plastic, or even a special electrolyte paste.

To begin, let us discuss in more detail how and for what purpose capacitors store energy. First, think about what happens when a constant voltage is connected to the terminals of our capacitor. As soon as a voltage is connected, electrons begin to flow from the capacitor plate connected to the positive voltage; electrons collect on the plate connected to the negative voltage. Eventually, equilibrium is reached, where no more electrons "flow." As that state is reached, the orbits of the atoms in the dielectric are elongated, and energy is stored much the way it is stored in a stretched spring. Why? The electrons in the atoms of the insulator are bound so tightly they cannot be freed. The dielectric, in conjunction with the plates, stores energy and provides the capacity for the storage and is how the name capacitor originated.

Even if we remove the original voltage, the voltage will remain the same magnitude as the source when we measure across the terminals of the capacitor. If a wire shorted the capacitor terminals together, a current would flow. In other words, the voltage cannot change instantaneously, but the current out can.

The current that flowed during that short-circuit would be a good indicator of the size of the *capacitance* or storage capability of the capacitor. The size of a capacitor is defined in *farads* or some unit of a farad. One farad is the capacitance for which a charge of 1 coulomb produces a 1-volt difference in the output terminals. Most capacitors are stamped with their value, but they come in many sizes and shapes. The size of a capacitor is often defined with a prefix. For example, many capacitors are labeled with the term "microfarads" rather than farads. A 1-farad capacitor is very large. A different capacitance is created by changing the plate area or making the distance between the plates different. In any regard, the stored electrons produce a voltage across the terminals of a capacitor. The bigger the capacitance, the more electrons are needed to produce a given voltage. Think of a capacitor like a piggy bank. Pretend the coins are the electrons stored in the electric field and the pile of the coins, the voltage.

The current flow from or into capacitors can change very quickly—in the ideal, instantly—but the voltage across their terminals cannot. The piggy bank is a useful analogy in this case too; when you add or subtract coins, it is just like current flowing in or out of the capacitor. The total number of coins, voltage, in this analogy, cannot change instantaneously. It is this characteristic that is used to make capacitors useful.

All this leads us to the standard capacitor equation used in calculating currents and voltages. This equation offers a concise look at what we have discussed:

current = (capacitance) x (change in voltage / change in time)

Can you see the relationship? One way of looking at this equation is that when the current into the capacitor is zero, the change in voltage over the change in time must be zero. This matches our piggy bank example. If you are not changing the number of coins, the total amount of money, or in our capacitor energy, must remain the same.

Another useful equation is about the size of a capacitance in relation to physical dimensions:

capacitance = (epsilon x plate area) / (distance between plates)

The epsilon is a constant, called the dielectric permittivity, and should be noted just to remember that dielectric characteristics affect capacitance. This equation also helps explain why two capacitors tied together in parallel results in a higher capacitance. You would be right in assuming that this would be associated to the plate area in the equation. And of course, the distance factor breaks down if the plates touch. That would be the ultimate breakdown.

Capacitors are often used to smooth ripple in DC power supplies to smooth out the voltage providing current as required. Because it takes a fixed and consistent time for the buildup of electrons to change the voltage on a capacitor from one voltage to another, capacitors are used in electronic timing circuits. Capacitors are also built with a fixed set of several plates connected to one terminal and a set of rotating plates that change the plate area by their position. Hence, we have the *variable capacitors* with which many radios are tuned.

Inductors

Inductors are more difficult to understand than capacitors. They are typically made of coiled wire, but we will begin by looking at a simple wire. As we know, a simple wire with current running through it will be surrounded by a magnetic field. So let us think about this wire as we change the voltage source supplying the current. As the voltage changes, the current change produces a changing magnetic field. The change in the magnetic field causes a voltage that resists the change in current. The resulting current resisting voltage is called a *counter-EMF*. Of course, when the changing magnetic field goes away, the counter-EMF also goes away, and to a DC voltage source, the ideal inductor becomes a short-circuit. The implication then is that a voltage cannot be present across the inductor when a counter-EMF is not present. In fact, the counter-EMF creates voltage. This process is known as *self-inductance* and by definition, involves the magnetic field that resists the changes in current.

Devices that demonstrate self-induction are known as *inductors*. A single wire has only a very tiny inductance, and that is why inductors are made of coils of wire. The effect is enhanced if the coil is wrapped on an iron core. Stop a minute and think about why this is true. The magnetic field is larger, and thus, the counter-EMF is larger. Inductors are also a type of energy-storage device because of the energy that creates the magnetic field.

The unit of inductance is called the *henry*. Joseph Henry was an ancient discoverer of electromagnetic induction. Officially, a henry exists when a current changing at 1 ampere/second produces an induced (counter) EMF of 1 volt. Does this make sense? Well, we have seen how a counter-EMF resists but does not stop a changing current. So we are told that we will measure inductance by the counter-EMF as well as the rate of change of current in the inductor circuit.

There is an equation to help remember how an inductor works that is comparable to the capacitor equation:

voltage = inductance *x* (change in current / change in time)

It is just like the one for the capacitor, but voltage and current are switched around. This equation tells us, when there is no voltage across the inductor, there can be no changing current over time. Do you understand why? If the current into the inductor is changing, there must be a voltage because there will be a counter-EMF.

Variable inductors are also commonplace. These inductors are made of a metallic core that can be moved into or out of the center of an inductor coil. The motion may be caused by a sliding mechanism, but it has also been accomplished by a threaded type of mechanism often with a screw head on one end.

Based on what we have discussed, you might understand how inductors are used in circuits that need their current slowed down during a start-up situation. Also, inductors are a major component of surge and spike protectors. Can you guess why? Surge protectors are devices that need to inhibit quick changes in current.

Transformers

Transformers also demonstrate some of the principles of self-inductance we've discussed. Another term for a transformer is *mutual inductor.*

Why are transformers defined as mutual inductors? Let us look at that single wire inductor again. However, now we will purposely lay another wire formed in a loop next to the wire attached to our voltage source. Furthermore, we will look at this new wire while a changing current is present, and the counter-EMF is being created in the first wire. The changing magnetic fields present in this situation induce a current flow in the second wire along with a resulting EMF. What we have described is a very simple transformer.

Since a changing current exists in the first wire, the *primary*, the resulting changing magnetic field *induces* a current flow in the second wire, the *secondary*. However, the flow of current in the second wire is in the opposite direction because of the counter-EMF. Of course, the total magnetic field is not coupled to the secondary wire, so the magnitude of the current and voltage in the secondary wire of this simple example is less than the primary.

Although we will expound on this further, it is worthwhile to examine the consequences of what we just learned. First, this is a simple example of how a radio transmitting antenna sends signals to the antenna of our radio at home. In this situation, the first wire is the transmitting antenna, and the second wire is the receiving antenna. And it also explains why the design of your TV is so important in helping prevent your local CB or HAM operator from interfering with your program. Wires everywhere can act like little primary or secondary coils in a transformer. When a primary creates an undesired current, we have *electromagnetic interference.*

As a demonstration of magnetic interference, put on the headphones to your battery-powered radio and turn it on. Listen to what happens when you turn your dust-buster on and off while they are close together; battery-powered items are foolproof examples. Did you hear the interference when the dust-buster was turned on? Note: Other things besides a dust-buster can be tried for this experiment. A whole field of engineering has been built upon measuring and eliminating electromagnetic interference.

That was an important sidetrack, but let us finish with transformers. Transformers are really made with coils of wire. To couple as much of the magnetic field as possible, the coils are often wound on iron cores of various shapes and sizes.

Now let us stop and think about the coils of the primary and secondary. For every loop of the coil, we can envision one wire that is either helping change the magnitude of the magnetic field for the primary or the magnitude of the current and voltage in the secondary. That is why the ratio of the turns or loops in the coils determines the ratio of the primary currents and voltages to the secondary currents and voltages.

This ratio business is an important thing to remember about primary and secondary coils. That is what allows us to have a secondary at a higher or lower voltage than the primary. When the voltage on the primary quits changing, the transformer quits because there is no magnetic field left. In fact, if a DC voltage is connected to most transformer primaries, it would almost be like connecting it to a short circuit.

The transformers with which many of us are familiar must operate with some sort of alternating current voltage source. This means they can be made in such a manner that the secondary is electrically isolated from the primary. This is great news! Step-down transformers (secondary voltage is less than the primary) used around the house make it possible to safely use low voltage created by the AC from our wall outlet when we use an appropriate transformer.

Summary Comments

Wire/insulation, resistors, capacitors, inductors, and transformers are the basic building blocks for electrical and electronic devices; electronic devices use semiconductors.

Combinations of these parts can lead to some practical and worthwhile voltage/current control situations. The capacitor and the inductor also allow us to store energy. The equations at the end of each section summarize and hopefully will help us remember how currents and voltages are related in each type of device. Wire and insulation are parts that, based on the fundamentals of electricity, make any electrical device possible. Transformers, we found, help us induce current from one circuit into another electrically isolated circuit. We also learned that the ratio between the number of coils in the primary and secondary of a transformer allows the voltage in the secondary to be different from the primary. And while beginning to learn about transformers, we learned that the same concepts explain radio waves and unwanted interference.

New Terms

CAPACITANCE - The current flow over time divided by the change in voltage.

CAPACITORS - Components made of a dielectric sandwiched between two conducting plates. They store energy in an electric field in the dielectric and as a result, resist changes in voltage.

ELECTROMAGNETIC INTERFERENCE - Unwanted mutual inductance effects.

FARADS - The measure of capacitance; 1 farad occurs when 1 ampere flowing over a second causes a change in voltage of 1 volt.

GAUGE - Indicates wire size; bigger numbers indicate smaller wires.

HENRY - The measure of inductance; 1 henry occurs when a voltage of 1 volt is produced when 1 ampere flows during a second.

INDUCTOR - A conductor element in which the energy stored in the wire resists changes in current. This process is enhanced in a coil of wire.

MUTUAL INDUCTOR - An inductor device that has a secondary coil electrically separate from the main primary coil in which current/voltage is induced. This is what makes a *transformer* work.

POTENTIOMETERS - Variable resistors in which one contact slides over the resistor material, usually carbon, to change the value of resistance.

POWER DISSIPATION - The energy dissipated in a resistor or other device, usually by heating itself up.

PRIMARY COIL - The coil of a transformer to which the voltage source circuit is connected.

PRINTED CIRCUIT BOARD - A way of manufacturing a circuit to more easily make it over and over.

RHEOSTAT - A device much like a potentiometer but bigger because it is made with wire as the resistor material.

SCHEMATIC - A way of describing a circuit with special symbols for each kind of part showing how they are connected.

SECONDARY COIL - The coil of a transformer that derives its voltage/current by induction from the primary.

SELF-INDUCTANCE - The voltage produced divided by the change in current over time.

CHAPTER 8

BASIC ELECTRIC MACHINES

Summary of the First Part of the Book

We have completed the part of this book that has until now dealt mainly with the how, what, and why questions concerning electricity. We have discovered how the wonderful thing called charge succumbs to the simple rules stated by Coulomb's Law to allow electrons to flow through some load to accomplish useful work. In the process of doing this, we have become aware of the basic terminology of electricity. We have also learned how devices are represented in a schematic.

In addition, we have learned how electricity relates to other important phenomena, such as magnetism and light. With that knowledge as a basis, we were able to investigate the major ways to create a voltage. The last chapter began an investigation into the most basic parts that allow the use of electricity.

It is in this second part that we learn about some real devices and machines. The first devices we call electrical machines: those devices that use the concepts of electromagnetism to cause a device to do something with mechanical movement. I have selected some major devices:

1. Bells
2. Relays
3. Solenoids
4. DC motors
5. AC motors

In addition, in later chapters, we will discuss how to control electrons using semiconductors for more difficult tasks. From our macroscopic view of the world, these applications do not necessarily involve mechanical motion. I will classify them as *electronic devices*.

Basic Electric Machines

Bells

One of the most basic electric machines is the simple electric bell. It is a device that makes a sound by the motion of a hammer striking the bell. The key to the motion of the hammer is an electromagnet with a switch in voltage circuit. The clever part is that the switch is part of an armature attached to the hammer. The electromagnet coil is positioned perpendicular to the armature. In electromechanical machinery, the part that moves is the *armature*. The basic sequence is as follows:

1. The switch is closed, causing the electromagnet to energize.

2. The magnetic field of the electromagnet pulls in the armature. This opens the switch, and the electromagnet begins to lose its magnetic field.
3. The momentum of the armature causes the armature to move the hammer and strike the bell.
4. The armature springs back and ends up in the position where the armature closes the switch again.

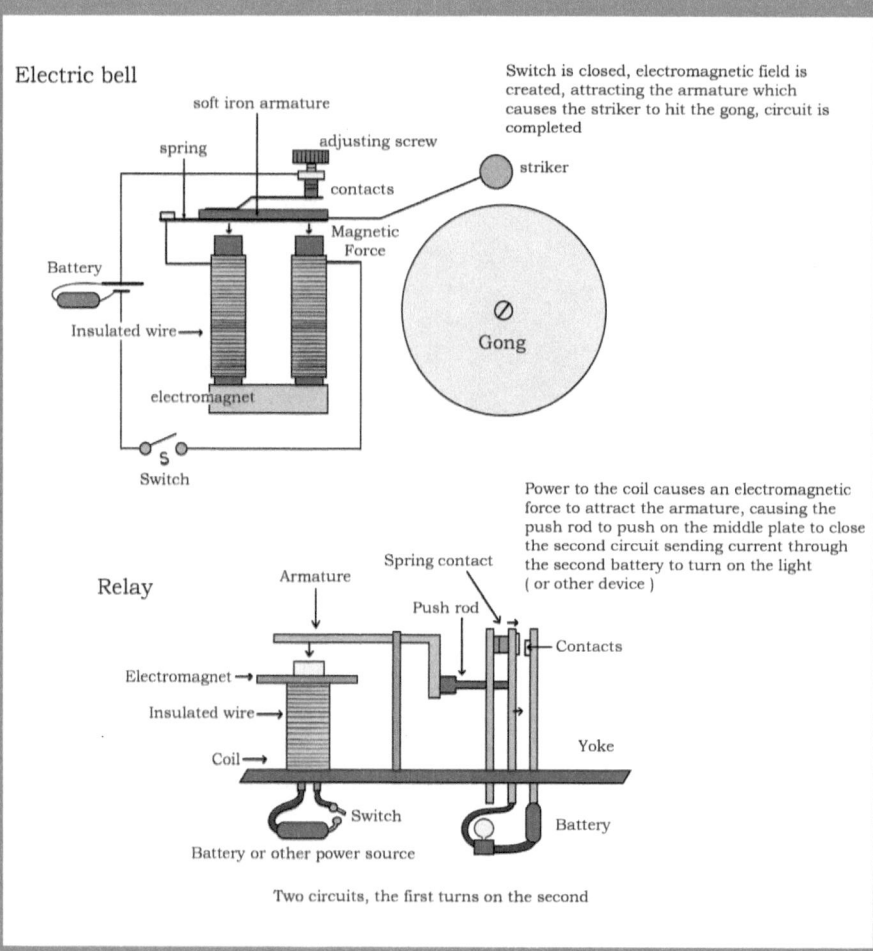

Relays

Relays are one of the first control elements used in electrical devices. Like a bell, they consist of an electromagnet and an armature, but in this case, the armature opens or closes contacts in a switch that is part of a separate electrical circuit. There are several important characteristics to know about relays. These are

1. Coil voltage - the electromagnet is designed so it will attract the armature at a certain voltage and current. The armature releases at another voltage, usually lower than the first because it takes less energy to hold the armature than to attract or "pick" it.
2. Contact rating - this is the current rating of the contacts on the switch that the armature opens and closes.
3. The switch has characteristics like a hand-operated switch. In other words, it has "throws" and "poles." For each wire circuit that can be controlled, a *pole* is defined. There are connections to be made when the armature is picked or released. This is defined by the relay *throw*. A standard control relay might be a double pole double throw (DPDT).
4. It is important to consider the position of the armature when a switch is closed. If it is closed when the coil is energized, then it is a normally open contact. Otherwise, it is normally closed contacted. Normal is the state before the coil is energized.

Solenoid

The solenoid is very similar to the relay, except the armature is a movable iron piece pulled into the open coil. The purpose of the solenoid is to translate the electrical control signal into some sort of mechanical motion. Usually, a spring is involved to reposition the armature after it has been pulled into the coil. The coil has similar rating considerations to the relay.

DC Motors

Just as we learned that generators provide either a DC or AC output, there is also a special type of motor to respond to each type of voltage. We can learn much about the workings of a DC motor by applying what we learned about DC generators. DC motors also have field windings, commutators, armatures, and brushes. There are many variations of the DC motor, most varying in how the field windings are connected or how the armature is wound. We won't spend a great deal of time trying to understand the different types of DC motors but will just understand the basic concepts.

As current flows through a wire or a coil, a magnetic field is established around the wire. In the case of a DC motor, a magnetic field is established in an armature that can turn or rotate. This spinning occurs inside the magnetic field of the field magnets. The polarity of the fields established in that armature is opposite on each end of the armature coil. If those fields of the armature and the fields of the field magnets are not parallel, a torque is established on the armature that tends to rotate the motor shaft. On each half turn, the brushes push current in an opposite direction because of the presence of the commutator, reversing the magnetic polarity of the armature so the magnetic forces will spin the shaft in the same direction for the whole revolution.

A simple DC motor has one drawback: There is one spot on the rotation, called the *neutral plane*, in which the fields are parallel. In this area, there is no magnetic force rotating the armature. However, if the shaft is turning, momentum carries the armature into the next torque area. This area is close to where the commutator switches armature polarity. If not for the commutator, the motor would end up vibrating in the neutral plane. Because of the presence of the neutral plane, sometimes it is not possible for the motor to start without manually rotating the shaft into a torque area. Also, these simple DC motors are not practical because of the uneven torque that results from this problem. In practical use, at least one additional armature is used, along with various schemes of brushes and winding connections.

In a motor, the armature is the load, and this load is some combination of wire coils that are very low in resistance. Current flows through the brushes from the motor terminals into the armature. The bad part is that this resistance is very low, about 1 ohm or less, which would normally cause a very high current. However, when the motor armature turns, even though the goal is to turn the shaft, the armature also generates its own *counter-EMF*. This counter-EMF causes the effective armature resistance presented to the power supply to be much higher. It's kind of unique in that, as a load slows the motor, the counter-EMF lowers, the armature resistance decreases, more current is drawn into the armature, and the motor speeds up. In fact, this simple motor sort of self-regulates its speed. Although other motor circuit arrangements are possible, this counter-EMF is a factor in all of them.

The circumstances involving the counter-EMF also would cause an extremely high current to flow, during starting before the counter-EMF builds up. This requires a starter circuit, which often has a built-in speed controller. The starter limits current, preventing damage to the brushes and commutator. Only DC motors can have a controller that limits speed because the speed is related to the strength of the magnetic field.

AC Motors

With commercial power generated as AC, it makes practical sense that many of the motors in use should operate using it. In the case of the basic AC motor, there are differences that set it apart from a DC motor.

In its most simple fashion, there is a wound field winding called a stator in AC motors, and a permanent magnet rotor, which is equivalent to an armature; there can also be a rotor coil supplied with DC through slip rings. As the AC in the stator changes, it just sort of pushes/pulls the rotor around. There is some amount of control required to get the rotor started and turning in the desired direction. For example, until the rotor is turning fast enough to keep up with the sixty-cycle AC, or *synch* up, it could not possibly run.

A way around these problems has been to cause the stator field to not just alternate but also rotate. That means, during start-up, the magnetic field of the stator will always be able to pull the rotor up to speed. This whole method is a little complicated for this book, but I am interested that you can visualize the concept and understand what sets an AC motor apart from its DC counterpart. The rotation is caused by using the special phases of the voltages connected to the stator winding.

The characteristic of the rotor sometimes requires an adjustment in the rotor during start-up to get the stator and rotor in synch. Often a capacitor is used to adjust the phase during start-up, often being automatically disconnected when *synchronous* speed is approached. It is synchronous because the moving magnetic fields of the stator and the rotor match.

Some manufacturers have created motors that run on AC and DC. In any such case involving DC, commutation is involved. That brings up the point that the AC motor described does not require commutation. This means AC motors lack some of the parts that often wear out.

Summary Comments

This section has covered some of the earliest basic electric machines. The use of motors spurned the industrial revolution, and such devices are now an important part of every home. We have taken a closer look at some basic machines, including the electric bell, the relay, the solenoid, and the DC motor. We have taken a brief look at what makes the AC motor different from the DC motor.

New Terms

COUNTER-EMF - This is the voltage generated during the rotation of a DC motor. The current produced flows in the opposite direction of the primary current flow into the armature, making the motor turn. This makes the motors practical by limiting the current flow and also helping regulate the speed.

NEUTRAL PLANE – The point in the rotation of the DC motor armature where its magnetic flux lines are in parallel with the magnetic flux lines of the field, causing no torque to be caused.

SYNCHRONOUS SPEED - The operating speed of an AC motor and the point at which the rotation of the armature matches the alterations in the magnetic field caused by the alternating current in the stator, like field coils.

CHAPTER 9

SEMICONDUCTOR DEVICES AND APPLICATIONS

We were introduced to semiconductors when we looked at solar cells. Now is a good time to look at some specific semiconductor devices used for control, storage, and display. There are a myriad of such devices, but I believe the subset selected will provide an idea of how semiconductors work.

P-N Junction Diode

This diode uses a P-N junction to accomplish a new task, as far as our learning goes. Quite often it is necessary to block current flow in one direction yet allow it in another. This device seemed to fit the bill just fine, and let's examine why. Here, we have a P-type material that tends to eliminate free electrons, holes, and an N-type material that tends to yield extra free electrons. They are produced into a single P-N junction by taking one crystal and injecting one impurity on one side and a different impurity on the other side. Terminals are fused onto each side, and BAM!—we have a *diode*. Of course, these days, to keep it clean, the tiny junction is enclosed in glass.

The way they work must be examined in two steps. First, we attach a battery, so the minus terminal is connected to the P-type material and the plus to the N-type material. In other words, the current or the electrons are trying to flow into the whole section and are readily taken in, while electrons flow out of the N-side into the positive terminal of the battery, *forward-biased*. On the other hand, if we try to force electrons through the N-type material and pull electrons out of the P-type material, we find the static barrier at the junction increases, and only a very tiny current flows, *reverse-biased*. Remember with any P-N junction, a natural barrier forms when the excess electrons from the N-type material fill the boundary holes. Then adding electrons to the N-type material only makes a larger barrier.

Diodes are required for rectifying AC, for example, when DC is needed and only AC power sources are available and must be converted to DC.

Semiconductor

A combo of elements that are usually insulators but when combined become conductors and become either (+) or (-) in a process called doping

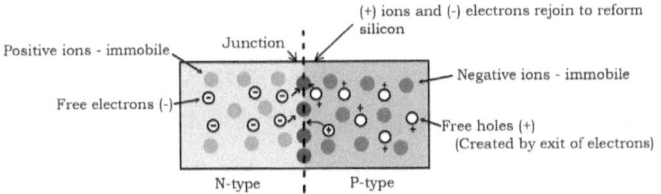

N-Type is often Silicon & Antimony (-)
P-Type is often Silicon & Boron (+)

No power present

Junction Diode

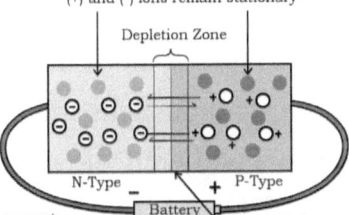

Electrons move from N to P as current
Holes move from P to N as current
Electrical current will only flow in one direction

Forward Bias

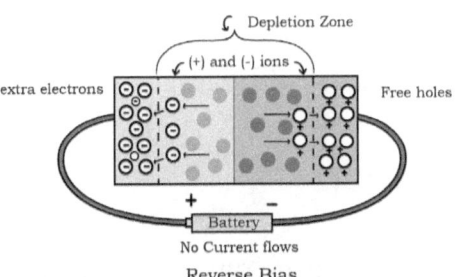

No Current flows

Reverse Bias

Also, diodes are used to detect radio signals and were made quite famous in "crystal radios," which use a crystal diode detector. The signal detector of a radio is shown below. In addition to the diode, a capacitor is needed to detect the superimposed voice waves. A radio wave is created by sort of shaping a high frequency, modulating, with the voice signal desired.

Later diodes became essential when they were used as a sort of current check valve to create digital logic circuits.

In computers, logic circuits often use two binary states; 0 represented by less than maybe 1.5 volts, and 1 represented by maybe 3–5 volts. These ranges may be different on real computers.

LEDs, Light Emitting Diodes

LEDs are a special sort of diode made with a semiconductor that causes the diode to emit light when forward-biased. Only certain semiconductors will emit the light. For example, silicon does not. In modern LEDs, we find uses in replacing incandescent lights and in the new flat-screen LED TV and even traffic lights. In TVs, the same signal used to control the electron gun provides information for an array of LEDs.

Bipolar Transistors

In the simplest of terms, one might imagine a *bipolar transistor* is two P-N junctions constructed of three pieces of semiconductors.

The three regions are referred to as the *collector* (C), *base* (B), and *emitter* (E). The B-E and the B-C represent junctions very much like the PN Junction we discussed earlier. In fact, for different transistor characteristics, the doping can be changed so the type of semiconductor in each region is reversed. For example, the "P" region can be changed to an "N" region. This change causes the current flow and the associated voltages required for the transistor operation to reverse.

These changes in doping result in two different types of transistors referred to as NPN or PNP. The names themselves show what kind of doping is used for each type. It also indicates the sandwich-type physical arrangement used to make a transistor. These two regions are the reason this kind of transistor we are looking at is referred to as a bipolar transistor. When the original transistor was made, it was a form of this bipolar transistor.

There are three separate circuits in this transistor; one operates through the base to the emitter, the second through the special collector emitter junction. The power supply circuits which make those two significant operate through the base-emitter junction and the base-collector junction.

This power circuit is known as the biasing scheme for the transistor. When the collector voltage is between the voltage source voltage and 0 (ground), the transistor is thought to be in the active range. When in the active range, a signal input to the base via the base emitter circuit can change the collector voltage level.

The magnitude of the collector emitter current is, within limits, controlled by the amount of current flowing in the B-E circuit. In fact, there is a certain factor of gain that multiplies the effect of the base current on the current flowing into the collector, assuming the transistor is in the active region. When the transistor is used to take advantage of this fact, the result can be an amplifier.

The same concept of holes and free electrons, as found in the diode, allows this transistor to function. The spacing of the junctions is critical, or the structure will really be two diodes, and the C-E current will not exist. Let's look at a PNP transistor. First, the EB junction will be forward-biased with electrons injected from the base and holes from the emitter. The C-B junction is reverse-biased, and because of the excess holes coming from the emitter, there is a tendency for holes to float to the collector; thus emitter and collector names originate. In fact, most of the holes reach the collector without recombining with a free electron. This happens only when the base width is very small. Now we need to look at some equations to make sense of this fact. First, the injected holes from the emitter, we will call IEH; the collected holes, we will call ICH; the electrons emitted at the base and going to the emitter, we will call IBE; the electrons that recombine with floating holes in the base, we will call IBEH; and ICBE is thermally generated electrons that drift from the collector to the base. In the following equations, we find IE, emitter current; IC, collector current; and IB, base current. Now:

$$IE = IEH + IBE$$
$$IC = ICH + ICBE$$
$$IB = IE - IC$$

It is defined that the current gain factor = A0 = (almost) ICH/IE, the ratio of the collector hole current to the total emitter current.

This definition leads us to the conclusion that

$$IE\,(A0) = ICH,\ IE\,(A0) = IC - ICBE$$

Since ICBE is very small, we can ignore it, and

$A0 = IC/IE$, the gain factor of the transistor.

This gain is a direct result of the actions of the holes and electrons by the semiconductor junctions formed.

Transistors can be designed into a circuit to operate as linear signal amplifiers or as a component in a digital logic element.

Other Semiconductor Devices

Other electronic devices include field effect transistors (FET) and silicon-controlled rectifiers or thyristors. We are not going to go into them in any detail in this book, but I encourage you to do further reading, if you are interested. I just wanted to expose you to their names in this book. They represent commonly used devices in amplifiers and control circuits. The thyristor is often used in power circuits.

CHAPTER 10

THE ORIGINS OF THE COMMERCIAL POWER SYSTEM

We have discussed much about what electricity is, how to generate it, and how to make devices that take advantage of its presence. One of the big questions might still be how and why electricity comes to our homes the way it does. Let's try to look at that very question.

Components of a Power System

Although it's often taken for granted, the electric power system is one of the marvels of modern society. This system consists of the following parts:

POWER-GENERATING DEVICES - As described in the previous chapter, devices that produce voltage and supply current, usually in large amounts.

TRANSMISSION NETWORKS - Long-distance senders of electricity; these lines carry electricity at a very high voltage to be cost-efficient.

DISTRIBUTION NETWORKS - Moderately high voltage feeder lines that distribute electricity from the substations, which terminate the transmission lines.

CONTROL MECHANISMS - Various instruments that control and route electricity; they also involve special monitoring and safety circuits.

LOADS - the end-users of the generated power; the variability of the loads makes the system even more remarkable since system designers have no real control over its magnitude.

Electric Lights: The First Load

Thomas Edison was the American inventor/entrepreneur who had an original vision for our power system. Although he is often cited for inventing the incandescent light bulb, he also deserves credit for many other devices used in the power system. We remember Edison because he didn't stop with the invention of one light bulb; instead, he originated an entire system of lighting.

Edison's technical and financial goals were to supply a lighting system to compete with gas lighting. And with his new incandescent bulbs, light was able to reach areas that would have been affected by a gas light's noxious fumes.

The Light Bulb's Inner Workings

To understand the rationale behind our power system, we need to take a closer look at the light bulb. The electric light, as simple as it seems, is the product of a lot of time and creative thought.

The major parts of the modern incandescent light bulb are the evacuated globe and the filament. The filament's characteristics were the key to making the first system a practical idea. The light bulb works because the electricity flowing through the filament causes it to glow.

To keep it from oxidizing and burning up, the filament had to be encapsulated in a vacuum. The next hurdle was finding a high-resistance durable filament. High resistance reduces current flow, making it easier on the generating system, which Edison needed to tie all the lights in parallel. A carbonized-paper filament was chosen for the first power system because it offered high resistance, long life, and an acceptable amount of light.

Today's light bulb has evolved from Edison's first lamp, and even some "newer" features of our current bulbs were developed by other early inventors.

The First Commercial Power System

Edison's original plan was to have electricity distributed from a central power station, not from individually owned generators. He predicted that, within a half-mile circle from the central plant, electricity could be sent to private homes and substituted for gas burners at a lower cost.

The lamps themselves greatly influenced the system. The goal was to have an electric lamp that produced an intensity of light equal to gas lamps. To provide the same lumens with the incandescent bulb's carbon-paper filament, the system needed to provide approximately 115 volts. When the switch was pulled on Edison's first power system, it powered 1,200 electric lamps by producing 850 amperes at around 115 volts.

This first generating system was an experiment to introduce the concept of electricity to a downtown area. The lamp was the start, but other pieces were essential. For example, a small device monitored the voltage and would signal an operator when to turn a wheel to change the amount of field resistance. This would vary the flux lines at the armature. The monitor was a simple, relay-like device that closed one pair of contacts at high voltage and another pair at low voltage.

Besides using such large parts as the generator and the massive distribution system, electricity brought the introduction of motors, which became another major system load.

The Three-Wire Distribution Scheme

Recognizing the importance of minimizing the transmission cost, Edison next invented the three-wire system of distribution.

In the original system, as the current increased, the cross-section of wire also had to increase to carry the flow without heat, which would have been a safety hazard. With the three-wire system, the center wire only carried the difference current of the two phases needed to supply the loads. This meant the system's wiring could be significantly reduced. In other words, the center wire could have a smaller diameter, so the system would cost less.

Also, more power could be delivered by using the generator voltages in series. This results in an 115V/230V system, and our large power users today, such as electric dryers and furnaces, take advantage of this 230V feature.

Why do we use AC instead of DC?

Edison was a proponent of DC power systems, although many of the system components he invented worked equally well for AC. Soon a big controversy developed over the merits of AC versus DC. Several technical and economic factors were behind many of the issues, and the debate was fueled by the spread of electric power to Europe.

When the transformer was introduced to convert AC power, systems could be designed to transmit very high voltages to remote locations, where the voltage was stepped down for use. This was very economical because high voltage meant the same power could be delivered to the load with a lower transmission current. Transmission costs could be less because the wires did not need to be so massive, and one generator could supply power to users at a greater distance.

Transformers are designed to convert voltages. For example, a low voltage, high current input can drive a load requiring a high voltage and low current. In an ideal transformer, voltage x current or power in and out are the same.

To take advantage of the transformer, AC power became the heavy favorite, especially when couplers were invented to convert generated DC to AC. A *coupler* is a short term for a system that converts one type of power into another, often using a motor-generator pair. Also, generators seemed to be simpler when made for AC rather than DC. The battle of AC over DC was fueled too by the invention of some very important types of AC motors.

The final decision was based on a compromise: Cities could stay with DC, but AC was necessary for providing electricity to rural areas far away from generators.

U.S. Power System Standards

As AC became the primary system, people soon realized some standards would be necessary, primarily for voltage and frequency. At that time, the common loads were incandescent lights, arc lights, induction motors, synchronous motors, and transformers. After much discussion, 60 hertz, cycles per second, became the compromise value for the generators as well as the loads, and the voltages and the three-wire system were taken directly from Edison's original power systems. Unfortunately, international agreements were not made, and the world is still encumbered with a variety of power systems.

Summary Comments

The power system we have now is a product of evolution and contention. Many of the decisions responsible for our current system were made for economic reasons, with technical compromises. In this chapter, we talked about why our power system, including the voltages, the frequency, and the standard three-wire system, exists the way they do.

The current power system, although not perfect, is still a remarkable engineering accomplishment. Without the insightful decisions made in its early days, electricity today might be considered a luxury rather than a necessity for life.

BIBLIOGRAPHY

Hughes, Thomas P. *Network of Power*. The John Hopkins University Press. Purpose: Research origin of US power system.

Sutton, Christina. *The Particle Connection*. Touchstone – Simon and Schuster, Inc. Purpose: Investigate particle interaction, when trying to define charge.

Mileaf, Harry, Editor in chief. *Electricity – Seven*, Revised Second Edition. Hayden Books. Purpose: Used to get perspectives on motors

Friedel, Robert and Israel, Paul with Finn, Bernard. *Edison's Electric Light*. Rutgers University Press.

Slaughbaugh and Parsons. *General Chemistry*. John Wiley and Sons.

Pierret, Robert F. *Modular Series on Solid State Devices*, Field Effect Transistors. Addison-Wesley Publishing Company.

Milnes, A. G. *Semiconductor Devices and Integrated Circuits*. Van Nostrand Reinhold Company.

INDEX

A

AC Generators 44-5, 51
AC motors 68, 74-6, 90
air 18
alternating current (AC) 39, 44-5, 50-1, 63, 74-5, 89-90
amperes 9, 12, 14-15, 88
armature 41-2, 44, 51, 68-9, 71-4, 76, 88
atomic spectra 34, 37
atoms 1-3, 5, 7-8, 12, 21, 24-5, 27, 31-2, 34-5, 37, 48-9, 56
  normal 2
Avalanche breakdown 22

B

base current 80-2
basic electric machines 68-9
battery ix-x, 11-12, 27-8, 45-7, 51, 78
  lead-acid 48
battery cells 46
Bipolar Transistors 80-1
brushes 41, 44, 51, 72-4

C

capacitance 57-8, 64
capacitor equation 58, 60
capacitors 56-9, 63-4, 80

carbon 46, 56
carbon materials 10
cells 46-7, 51
charges 3-5, 7, 10-11, 20, 45, 47, 57, 93
chemical reactions 20, 46-7
coil voltage 71
coils 41-2, 44-5, 51, 59, 62-3, 71-2
collector voltage 80-2
colors 1, 34-7, 48
commutator 41-2, 44, 51, 72-4
conductivity 11, 17, 21, 48
conductors 10, 12, 15, 17, 21, 36, 41, 44, 46, 50, 52-5
contact rating 71
control mechanisms 86
coulomb 7, 9, 57
Coulomb's Law 7-9, 67
counter-EMF 59-61, 73-4, 76
current 15, 18, 25, 57
current flow 10-12, 18, 24-5, 36, 41, 57, 61, 73, 76-8, 81, 87
cycles 39-40, 50, 90
cycles per second (CPS) 40, 90

D

DC generators 41-2, 44, 51, 72
DC motors 68, 72-6
dielectric 18, 22, 56
dielectric permittivity 58

diodes 78, 80, 82
direct current (DC) 39, 42, 44, 72, 74-5, 78, 89-90
distribution networks 86
doping 49, 52, 81
double pole double throw (DPDT) 71
dry cells 47

E

Edison, Thomas 31, 86-90, 93
electric bells 68-9, 71, 75
electric charge 3-4, 7
electric circuit 36-7
electric field 7, 9, 29, 57
electric lighting devices 35
electric lights 86-7
electric power 14-15, 89
Electrical Attraction 5
electrical machines 28, 67
electricity ix-xi, 1, 3-4, 14-15, 17, 20, 23-4, 30-1, 36-7, 45, 48, 67, 85, 87-8, 90-1
   static 10-12, 16
electrode materials 46
electrodes 20, 22, 46-7, 52
electrolysis 20, 22
electrolyte 20, 22, 46-7, 52
electromagnet coil 68
Electromagnetic Force 4
electromagnetic interference 62, 64
electromagnetic theory 28-30
electromagnetic waves 18, 28-30
electromagnetism 27-8, 30, 67, 69, 71
electron gun 35-6, 80
electron orbits 25
electronic components x, 60, 83
electronics ix, xi
electrons 1-3, 5, 7-10, 15, 17-18, 20-1, 23-5, 27, 31-2, 34-7, 48-9, 52, 55-8, 78, 82-3
   flow of 9-10

element 8
emitter 80-2
energy 5, 8-9, 14, 18, 28-30, 32, 34-5, 46, 48, 56
energy levels 3, 8, 31-2, 34
epsilon 58

F

farads 57, 64
field 18, 22-3, 25, 29, 32, 44, 72-3
field effect transistors (FET) 83, 93
field windings 44, 72
fluorescent light 35
free electrons 5, 48-9, 77, 82
frequency 34, 37, 39, 50, 91

G

gas 18
gas phase 18
gauge 54, 64
good insulators 10, 20, 48

H

henry 60, 64
Henry, Joseph 60
Hertz, Heinrich 40
holes 49, 52, 77, 82-3
hydrogen 34, 46

I

ice 17
inductors 59-60, 63-4
   mutual 61, 64 *see also* transformers
insulation 53, 63
insulators 10, 12, 15, 17-18, 21, 53-6
ions 18, 20, 46-7
   negative 5, 8
   positive 5, 8

J

junctions 77-8, 80, 82

K

Kilowatt, Reddy ix

L

lamps 87-8
leakage current 10-11, 15
light 15, 31-2, 34-7, 48-9, 67, 80, 86-8
light bulb xi, 36, 86-7
light-emitting diodes (LEDs) 80
light waves 28-9
lightning 18
Liquid Phase 20
lithium battery 48
loads 86, 88-90
logic circuits 80
loops 25, 27, 42, 61-2

M

magnetic field 23, 25, 27-30, 41, 45, 59-60, 62, 69, 72, 74, 76
magnetic flux 25, 30, 41-2
magnetic flux lines 50-1, 76
magnetic interference 62
magnetism 15, 23-4, 27, 30, 36, 67
magnets 23, 25, 27
    permanent 27
matter 1, 3-4, 7-8, 24
Maxwell's equations 29-30
metal wire 10
microfarads 57

N

N-type material 49, 52, 77-8
neutral plane 73, 76
neutron 2, 7-8
nickel-cadmium combination 48
nucleus 2-3, 5, 8, 24-5, 31

O

Ohm 15
Ohm's Law 12, 16
oil 20
oscilloscope 40, 51

P

P-N junction diode 77
P-type material 49, 52, 77-8
particles 1-3, 7-8, 23, 29, 31
    messenger 4-5
period 51
periodic table 3, 8, 48
phases 17, 44, 75
    physical 17
photo-voltaic cell 49
photons 29, 31, 34-5, 48-9
    virtual 4-5
piggy bank 57
Planck, Max 32
Planck's equation 32, 34, 37
polarity 28, 41, 44, 72-3
potential voltage 15, 45
potentiometer 56, 65
power 14, 45, 55-6, 90, 93
power circuits 81, 83
power dissipation 55, 65
power-generating devices 85
power system 85-8, 90-1
  first commercial 87
primary cells 46-7, 52
primary coils 64-5
Printed circuit boards 54, 65
protons 1-3, 7-8, 31

Q

quantum energy 32, 34
quantum mechanics 32, 37

R

radiation, released 31
radio   ix-x, 24, 59, 61, 80
radio waves   28, 40, 63, 80
relays   68, 71-2, 75
resistors   10, 16-17, 55-6
rheostat   56, 65
rotor   44, 51, 74-5

S

schematic   53, 65, 67
secondary cells   46, 52
secondary coils   62, 64-5
self-inductance   59, 61, 65
semiconductor devices   77-83
semiconductors   48-9, 52, 63, 68, 77, 80-1
silicon   48-9, 80
simple wire   59
simplest atom   34
single wire   59
solar cells   48-9, 52, 77
solenoid   68, 72, 75
Solid Phase   20-1
solid wire   54
solution   20, 22, 46
sparks   18
states, liquid   20
stator   44, 51, 74-6
storage cells   46-7
stranded wire   54
sulfate ions   46
switch   68-9, 71, 88
synchronous speed   75-6

T

Three-Wire Distribution Scheme   88
thunderstorm   18
transformers   61-3, 89
transistors   xi, 81-3
transmission networks   85

V

valence electrons   21-2, 49
valence number   3, 8, 48, 52
voltage   11-12, 14, 16, 18, 36, 39-42, 44-7, 49-51, 54-63, 67, 72, 75-6, 88-91
   induced   41, 50
voltage generation   50
Voltage Induction   41
Voltage Sources   39, 59, 61, 63
volts   11-12, 14, 16, 88

W

water   17
watts   14, 16
wire   53-4, 57

Z

zinc   46-7
zinc electrodes   46-7

www.ingramcontent.com/pod-product-compliance
Lightning Source LLC
Chambersburg PA
CBHW020443220526
45464CB00002B/827